Management for Professionals

More information about this series at http://www.springer.com/series/10101

Wolfgang Lehmacher

The Global Supply Chain

How Technology and Circular Thinking
Transform Our Future

 Springer

Wolfgang Lehmacher
World Economics Forum
New York, USA

Translation from the German language edition: Globale Supply Chain, by Wolfgang Lehmacher © 2016 Springer Fachmedien Wiesbaden. Springer Fachmedien Wiesbaden is a part of Springer Science + Business Media. Translation: Ann-Christin Wimber, ras kommunikation.

ISSN 2192-8096   ISSN 2192-810X   (electronic)
Management for Professionals
ISBN 978-3-319-84570-8   ISBN 978-3-319-51115-3   (eBook)
DOI 10.1007/978-3-319-51115-3

Printed on acid-free paper

This Springer imprint is published by Springer Nature
The registered company is Springer International Publishing AG
The registered company address is: Gewerbestrasse 11, 6330 Cham, Switzerland

Foreword

Wolfgang Lehmacher's book is perhaps the most extensive and balanced treatment of the pervasive role supply chains play in our global economy and even our daily lives. Through exhaustive research into all aspects of the world system, he demonstrates beyond any doubt the extent to which an integrated world benefits all of humanity. Lehmacher has held key roles in supply chain with public and private companies all over the world, where he has been involved in major change and expansion projects – all of which give him a unique authority to produce the analysis presented here.

Importantly, Lehmacher stresses the combination of trade and logistics as twin pillars of openness. In one sense, this is the theory and practice of achieving integration of talk and action. Very few appreciate the organizations and bodies that, behind the scenes, enable the critical flow of goods and commerce that smooth supply chain functioning, lower costs, create jobs, and boost economies. Another great strength of this volume is the way it leads the reader through the full life cycle of the supply chain, thereby addressing concerns about ecological costs both upstream and downstream, from the water consumption required to produce T-shirts to local externalities of the mining industry.

The reader will also come to understand the role of exogenous and strategic variables that affect global supply chains, from sanctions due to geopolitical rivalry to protectionism owing to industrial policy. Emerging markets take center stage either as victims of volatility or through their ability to seize greater portions of the value added in high-tech supply chains.

The supply chain does not disappear in a digital world. Lehmacher also presents the state of the art in e commerce trends worldwide and dissects the growing importance of the "sharing economy" by which digital platforms have enabled radical shifts in consumer behavior and thus disrupted entire industries, from mobility to real estate to retail. The supply chain, as Lehmacher rightly points out, constantly shifts with the times, evolving as we do into a circular society.

Senior Fellow   Dr. Parag Khanna
Lee Kuan Yew School of Public Policy
Singapore

Preface

This book was first published in a German edition. Since then, much has changed in the global political and economic landscape. Britain has voted to exit the European Union, and the growth of Islamic State of Iraq and Syria (ISIS), also known by its Arabic language acronym Daesh, has created a refugee crisis facing the entire world, to name only two current international challenges. For this reason, I have amended portions of the book for the English edition to address the changing times. I remain convinced, though, that open borders and the circular economy, as long as they are paired with strategic foresight and socially accepted income distribution, are necessary to ensure wealth and prosperity for all.

I would like to acknowledge the help of Parag Khanna for writing the foreword; Ann-Christin Wimber for translating; Peter Lyons, Luisa de Miranda, Victor Padilla-Taylor, and Heike Steinmetz for reading and giving thoughts; and Springer for the opportunity to translate and update this book.

New York, NY, USA   Wolfgang Lehmacher
October 2016

Contents

Introduction

We are living in a small world. Buying raw materials in Africa and producing products in China, Mexico, or Vietnam for distribution and sale in Europe or the USA no longer seem like a big deal. Without effort, all manners of goods are crossing rivers and mountains, borders, and oceans – or so it appears.

In reality, the world is not that flat. Even though today's supply chain spreads across the globe, there are still many challenges and obstacles. Therefore, it is not surprising that one of the most respected global companies is led by a supply chain expert – Tim Cook, CEO of Apple.

The free flow of goods, services, capital, labor, data, and culture is by no means a given, not even in the twenty-first century. Pressured by misinformation, economic protectionism, and rising antiglobalization movements, national governments regularly erect new barriers to trade and disconnect their national economies from global platforms in the world of procurement, production, distribution, and repurposing of goods. In addition, national governments or economic blocs occasionally exclude other countries partly or entirely from the free exchange of goods by the imposition of trade sanctions.

Even countries operating in the same globalized and fluid economic environment, with the same trade agreements, tariffs, and barriers, sometimes take different routes. While Poland experienced a 23% increase in wages between 2007 and 2015, German wages grew by 14% and French wages grew by 10%. With a 10.4% loss in average wages in the same period, the United Kingdom (UK) has reached the same level of negative wage growth as Greece.

With citizens concerned about the loss of jobs and the potential for terrorist attacks, agreements such as the Trans-Pacific Partnership (TPP) and the Transatlantic Trade and Investment Partnership (TTIP) are facing strong political and societal headwinds. Even existing regional trade blocs, such as the European Union (EU), are experiencing tendencies towards greater disintegration – Brexit is one example.

"In aggregate terms, the human race has never had it so good. Life expectancy has risen by more in the past 50 years than in the previous 1,000. When the Berlin Wall fell, two-fifths of humanity lived in extreme poverty. Now it's one-eighth," write Ian Goldin and Chris Kutarna of Oxford University (Kutarna, 2016). But some have been left behind. Governments need not only to open borders but to address the shortcomings of globalization and create new opportunities for those negatively affected by the connected world. All positive effects that trade, the global supply chain, and globalization have produced mean nothing to those individuals. Governments must be aware of this fact. The world needs fair distribution of the growth and wealth created. Best positioned in this respect are economically strong and competitive countries with strong political leaders. Globalization can bolster economic strength through investment, knowledge transfer, capacity building, and competitive pressure and thus lay the foundations for a more inclusive world.

As a whole, the global economy has gained substantially from trade and globalization. And the supply chain is the heart of modern business and life. The "magic conveyer belt" carries raw materials and parts, seeds and fertilizers, and equipment and machinery to the factories and farms and any kind of product to shops and to the homes, offices, collection points, lockers, and boxes of individuals.

Advanced supply chain providers, such as logistics and transportation companies, offer more than traditional handling, storage, and movement of goods; some provide assembling services and 3D printing, for example. The lines between manufacturing, supply chain services, and retail are increasingly blurring. The supply chain and the value chain are basically inseparable. The "magic conveyer belt" can be considered as modern production and the supply chain as the new retail; logistics and transportation companies, such as SF Express in China or DHL, have already entered the e-commerce domain.

With many of the public not fully aware of the importance, scale, and scope of global supply and value chains, government support for international trade is often controversial. Indeed, national election cycles in recent years have yielded an increase in populist voices opposing the principles of globalization, trade, and foreign investment. While the external message of governments is usually in support of free trade and exchange, internally, governments are pressured to provide support and protection on behalf of vested domestic interests. While tariffs are in decline, nontariff measures (NTM) such as licenses and quotas are on the rise. The ability to access markets depends on the ability to overcome NTMs, i.e., increasingly by compliance with sanitary and phytosanitary measures, technical regulations, norms, etc. "The exact impact of NTMs on trade flows is not very well understood," writes the United Nations Conference on Trade and Development (UNCTAD, 2016). However, a holistic understanding of the supply chain system is the prerequisite for realizing its full potential and mitigating risks and undesired developments.

I hope to provide more clarity on the role of the global supply and value chain in sustainably elevating the prosperity and well-being of all by elucidating: (1) the need for fluid and seamless exchange as the basis for economic growth and wealth creation and (2) the necessity to transform the current, largely linear supply chain into a regenerative system to ensure the well-being of the global society. This book is an overview and analysis of the current state of the supply chain and directed at everyone interested in the world of supply and value chains, in particular chief supply chain officers (CSCOs) and senior supply chain practitioners in the public and private sector.

Global supply chain growth is an opportunity and strategic tool, which can produce economic growth and jobs and also solve other pressing challenges of our time. However, this requires a radical redesign. It is my intention, through practical examples, to provide a comprehensive overview of current societal and economic challenges as well as potential regenerative solutions enabled by circular applications of the global supply and value chain.

This publication should add more evidence that the global supply chain is beneficial for economy and society, while the design and performance of the supply chain are crucial for growth and wealth, prosperity, and well-being of nations. Therefore, I must touch lightly on the issue of the distribution of wealth – one key cause of the antiglobalization discussion and today's increasing disintegration tendencies. Governments have few options but to focus on driving economic growth and fighting income inequality, before and after tax.

The global supply chain is not a secondary function of growth and progress, but a primary enabler of the modern world. However, the importance of the global supply chain for economic and societal development does not demonstrate itself easily. Only with close scrutiny, we discover that the design and management of the global supply chain determine the economic strength and wealth of nations and, to a large extent, the future of mankind as a whole. Comprehensive assessment of the supply chain will highlight current obstacles to trade between nations. Deeper involvement with the topic will show the need for improvements in multiple areas, in particular in infrastructure, information and communication technology (ICT), education and training, policy decisions, legal frameworks, and governmental programs. Today, the tension is apparent between wealth creation and wealth distribution, as well as economic growth and the well-being of human beings and the planet.

The magnitude of environmental burdens, resource scarcity, and the potential for increased social unrest provide a good deal of urgency to the need to transform global supply chains towards an ecosystem that is balanced and regenerative. This reshaped ecosystem is called the circular supply chain, the backbone of the so-called circular economy and enabler of a cleaner, more secure, safer, and more inclusive future. The purpose of the circular supply chain ecosystem is to yield economic growth while improving the efficiency of resource utilization and protecting the environment. Everything which wastes resources should be eliminated. Understanding possible ways to build this ecosystem requires knowledge building and co-creation. We need new designs, new products, and new ways of producing and moving goods. We need ways to better monitor the use of products. We need to repurpose, i.e., reuse, remake, or recycle, out-of-use products.

All of us, all nations and all individuals, are involved in the global supply and value chain; we benefit from the "magic conveyer belt" and contribute to the system. Collectively, we are responsible for its design and its positive and negative outputs. Many of our actions affect people and the planet. Ideally, the global supply and value chain is the result of well-aligned and like-minded parties, often from different parts of the world, cooperating and collaborating closely to produce the perfect bundle of outputs required by modern and responsible societies – like an orchestra, which cannot delight the audience unless all musicians are playing in perfect harmony, but sometimes allows creative variations and improvisations. The ideal set of products and services is designed for zero waste, produced with renewable energy and zero emissions, accessible to everyone, and yielding well-distributed wealth for individuals, nations, and the global society.

We are all part of the orchestra called the global supply chain. We are the ones that actively shape every single process; we are supplier, retailer, manufacturer, farmer, distributor, consumer, reuser, remanufacturer, or recycler. Our wishes, intentions, and behaviors define, collectively, how goods and services are designed, produced, distributed, used, maintained, and repaired; we are the ones who also determine what happens after a product's first and last use cycles, whether the product or parts of it are reused or the resources are just wasted or "shelved" in landfills. Awareness of modern technology and practices and close collaboration among all parties along the supply and value chain are prerequisites for building the regenerative supply and value ecosystem.

Policymakers also have an essential role. High-performance supply chains require supportive frameworks, programs, and incentives favoring circular applications, including soft and hard infrastructure and agreements with other nations. The topic is complex. Therefore, all stakeholders need to understand the functionality, dynamics, and unwanted consequences along the global supply and value chain and the impact on profit, people, and planet.

New times require new concepts. The world we are living in is changing quickly, reinventing itself constantly through new technologies, new models, new players, and new ways to work and live and changing geopolitical balances and situations. The old gives way to the new, which sometimes causes pain and fear. With population growth and demographic imbalances; the economic power shift from the West to the East; climate change; natural disasters; financial, currency, and economic crises; political instability and social upheaval; refugees; poverty and famine; and terrorism, attacks, and wars, solving today's global challenges in our interconnected and interdependent world requires holistic approaches and systemic measures.

It is my aim to provide a better understanding of the barriers and enablers of a regenerative global supply and value chain ecosystem and to assess the current status of the supply chain, including progress towards achieving a seamless, fluid, and regenerative flow of goods. The analysis will explore various measures and best practices already in place and shed light on areas which require attention. The journey towards a more fluid and inclusive exchange of goods, without resource, environmental, or societal burden, is not a simple one.

Literature

Kutarna, C. (2016, June 8). *There's never been a better time to be alive. So why the globalization backlash?* World Economic Forum. Accessed September 14, 2016, from https://www. weforum.org/agenda/2016/07/there-s-never-been-a-better-time-to-be-alive-so-why-the-global ization-backlash

United Nations Conference on Trade and Development (UNCTAD). (2016, April 14). *New database of ASEAN non-tariff measures.* Accessed September 14, 2016, from http://unctad.org/en/pages/ newsdetails.aspx?OriginalVersionID=1234&Sitemap_x0020_Taxonomy=UNCTAD%20Home

Trade and the Global Supply Chain

Global trade is having a hard time not only through recent tendencies towards rising nationalism, stronger protectionism, and regional and global disintegration risks. According to the 18th Global Trade Alert Report by the London-based Centre for Economic Policy Research (CEPR), governments all over the world imposed 539 trade restrictions in the first 10 months of 2015, of which the G20 were responsible for 443 (Evenett and Fritz 2015). In comparison, 141 trade facilitation measures, i.e., supportive actions, have been established in the same period.

This imbalance sets the business world thinking, because trade is not only important for prosperity but also for peace. Matthew Jackson and Stephen Nei, both of the Department of Economics, Stanford University, Stanford, California, point out in their study, "Networks of Military Alliances, Wars and International Trade," that with an absence of international trade, no network of alliances is peaceful and stable. Military alliances, when combined with isolationist economics, they postulate, are no guarantee for lasting peace. Countries that have high levels of trade with their allies are less likely to be involved in wars with other countries. Based on historical data on wars and trade, Jackson and Nei showed that a country held on average 2.5 alliances between 1816 and 1950. There is only a 0.695 probability that an alliance will still exist after 5 years. Between 1951 and 2003, the number of alliances quadrupled to 10.5. At the same time, the probability of a collapse of the alliance dropped to 5%. Additionally, the scientists found that with the increase of trade, wars greatly subsided. Between 1820 and 1959, trading partners averaged 0.00056 wars per year, while from 1960 to 2000, the average was 0.00005 wars per year, meaning the factor decreased to less than one-tenth as much (Jackson and Nei 2015).

In recent years, military conflicts seem to occur more often: attacks around the world, directed or inspired by terrorists groups such as Islamic State (IS or ISIS),[1] and conflict in Eastern Ukraine, Afghanistan, and the South China Sea. These

---

[1]Hitherto referred to as Daesh

© Springer International Publishing AG 2017
W. Lehmacher, *The Global Supply Chain*, Management for Professionals,
DOI 10.1007/978-3-319-51115-3_1

events create stress on our economic system and fuel the antiglobalization argument and impact supply chains around the world and thus global wealth. Businesses big and small rarely manage without commodities, intermediates, parts, or machinery from other parts of the world. In the USA, for example, 56% of imports are intermediate products (US Chamber of Commerce 2015). Manufacturers depend on trade moving goods efficiently through the supply chain to remain competitive, to create jobs, and to promote economic growth. More than 28% of the US gross domestic product (GDP) is tied to trade. The US Chamber of Commerce predicts that this number will continue to grow.

At the same time, the fragility of the supply chain is evident. Earthquakes, volcanic eruptions, tsunamis, or simply a fire in a manufacturing plant can greatly impair the flow. As a result, relief supplies may not reach their destination; grocery stores and consumers will not get the product they have ordered; production lines could come to a standstill. Chief supply chain officers are responsible for mitigating supply chain disruptions. But this task also falls to governments.

A word on the definition of the supply chain, one of the commonly cited definitions about the nature of the supply chain, was given by Martin Christopher, Emeritus Professor of Marketing and Logistics at Cranfield School of Management in Bedfordshire, England. He suggests that the supply chain is "the network of organizations that are involved, through upstream and downstream linkages, in the different processes and activities that produce value in the form of products and services in the hands of the ultimate consumer" (Christopher and Peck 2004), upstream pointing towards the supplier end of the chain and downstream to the customer end. Transportation and logistics companies, manufacturers, suppliers, retailers, and distributors as well as consumers, re-users, remanufacturers, and recyclers are all parts and stakeholders of the system.

Another frequently used term is value chain, which describes all organized units of operation and process that create value and consume resources. The concept was introduced for the first time by Michael E. Porter in 1985. In his book *Competitive Advantage*, he states:

> The idea of the value chain is based on the process view of organizations, the idea of seeing a manufacturing (or service) organization as a system, made up of subsystems each with inputs, transformation processes and outputs. Inputs, transformation processes, and outputs involve the acquisition and consumption of resources—money, labor, materials, equipment, buildings, land, administration and management. How value chain activities are carried out determines costs and affects profits.

Stakeholders in the value chain are the same as the aforementioned participants in the supply chain system. Supply chain and value chain are intertwined. In today's world, none can meaningfully exist without the other. Therefore, both terms – supply chain and value chain – will be used interchangeably.

1.1   The State of the Global Flow of Goods

A world without a global marketplace seems inconceivable. LCD monitors from Japan, furniture manufactured in Sweden, kitchen appliances designed in Germany, and cosmetics made in France, products from all over the world, find their way into shops and homes all over the planet.

Smartphones are used by consumers all over the world. According to data from the International Data Corporation (IDC), the second quarter of 2015 saw 341.5 million shipments of mobile phones worldwide (IDC 2016). Mass production in today's global procurement, production, and distribution networks results in affordable electronics, considered luxury products not long ago. One example is Apple: product is designed and engineered in California, with the component parts fabricated and assembled elsewhere.

In the month of November 2015 alone, 190.5 million cell phones were produced in China (Statista 2016). High-quality mobile devices are comprised of approximately 40% metals, mostly copper, and 40% plastics by weight, with the remainder made up of glass and/or ceramic (Underwriters Laboratories Inc. 2011). Most of the metals come from mines in developing countries. Ninety percent of rare earth minerals, naturally occurring solids whose combination comprises essential smartphone parts, are mined in Mongolia. Displays and batteries come from Japan and Taiwan, while semiconductors can be sourced elsewhere: for example, the iPhone 5 semiconductor is supplied by STMicroelectronics in Geneva, Switzerland.

Globalization and global trade have influenced our lives in ways we simply don't even think about anymore (Evenett and Fritz 2015). Globalization is a matter of fact. An unobstructed global marketplace is widely understood to be a fait accompli.

Over the long term, as goods have gotten cheaper, wages have generally been increasing. In 1960, employees had to work 338 h in order to buy a TV; today they only have to spend 28 h at work in order to be able to afford a state-of-the-art model. The fact that appliances are getting cheaper is closely related to globalization. Globalization makes it possible for goods to be manufactured overseas, where wages and labor costs are lower. Globalization is not only about manufacturing but also about sales. Thanks to global distribution networks, goods can be transported and distributed almost everywhere. Serving the global market requires mass production, which brings cost reductions through economies of scale. Consumers know that electronic appliances are produced in China and clothes in Bangladesh. This has created the myth of the seamless supply chain and free trade. However, global trade is not as fluid as one might think. There exist import and export duties, quotas, licenses, sanitary and phytosanitary measures, technical regulations, and norms, which are meant to protect citizens, crops and animal populations, and also local industries and jobs.

After many years of significant progress towards the goal of more seamless international flows of goods, recently, many new barriers are emerging. The British Centre for Economic Policy Research has found an increase in trade barriers and asks in its 18th GTA report whether "The Tide Turns" (Evenett and Fritz 2015). The supply and value chain is often blocked or slowed by duties, taxes, tolls, labor protection laws, and other regulations. These arise when governments react to fears about health, security, or the potential loss of jobs. In both society and politics, we see strong advocates of free trade who believe in its positive effects. On the other side, opponents call for protectionist measures to be introduced by the government.

In Europe, data protectionists fear that personal data of customers may be vulnerable and unlawfully accessible on global Internet platforms, a fear customers do not seem to share when downloading applications on their mobile devices or joining social networks such as Facebook, Instagram, or WhatsApp. Another fear often troubling the global community is the potential risk for open global supply chains to inadvertently allow the introduction of invasive plant or animal species to an ecosystem. In addition, free trade may bring products regarded as undesirable or unwanted from a societal point of view into the country. Illegal products such as smuggled cigarettes or weapons may cross national borders. As a result, governments regulate the flow of goods by imposing laws, legal requirements, and testing specifications.

In the USA, the Trade Act of 1974, which was amended last in June 2015, sets the framework for foreign trade including trade agreements, safeguard procedures, treatment of high-technology products, agreements on nontariff barriers, and resolutions approving commercial agreement (Trade Act 1974). Like a lot of countries, the USA has established free trade zones (FTZ), geographic areas declared to be outside the normal customs regime of a territory. This means that, for foreign merchandise entering FTZs and re-exported as different products, customs procedures are streamlined and tariffs do not apply (Bolle and Williams 2013).

After many years of successful trade facilitation, new barriers have been established across the world, largely resulting from political conflict or socioeconomic tensions. One example of a complex political relationship affecting trade is the relationship between the USA and Russia. In 2014, the USA sanctioned Russia for its military actions in Ukraine, with limited impact. US exports to Russia have not declined massively. Between the second quarter and the end of the year 2014, US exports totaled \$8.94 billion[2] compared to \$9.4 billion in 2013 (Rapoza 2015).

A different picture emerges when looking at the European Union (EU). In response to the Crimea issue, the EU imposed trade-restrictive measures in July 2014, reinforced in September 2014. The actions included a ban on imports of goods originating in Crimea or Sevastopol, unless they have Ukrainian certificates. Furthermore, EU export restrictions prohibited the export of goods and services to Crimea in the telecommunications, transport, and oil/gas/mineral exploration sectors. The EU has frozen assets, withdrawn capital, and prohibited EU nationals and companies from buying or selling new bonds, equity, or similar financial

---

[2]All figures are US dollars on a nominal basis, not seasonally adjusted unless otherwise specified.

instruments from Russia (EU 2015). In the first quarter of 2014 alone, $70 billion worth of EU capital was withdrawn.

Russia's reaction to USA and EU measures hurt both economies equally. In the EU, German exports to Russia totaled EUR 38 billion ($51 billion) in 2013 – the highest in the EU. Germany gets more than 30% of its oil and gas from Russia. Italy and some of Russia's former Soviet bloc neighbors are highly dependent on Russian energy and gas deliveries (BBC 2014). Russia turned to Asia to sell their gas. In May 2014, it entered into a contract with China and settled on supplying its neighbor; China agreed to buy 38 billion cubic meters of gas per year beginning in 2018 and continuing for at least 30 years. Another reaction to the EU and USA sanctions was Russia's decision to stop the import of meat, vegetables, fruits, and milk from the USA, Canada, Switzerland, and Japan. To give tit for tat, they applied visa bans to European politicians, just as the EU had banned 139 Russian persons from crossing the border.

German mechanical engineering suffered from the restrictions. Instead of contracting German suppliers, Russia now favored domestic businesses. Germany used to be one of the most important foreign investors in Russia. In 2012, the accumulated German direct investments added up to EUR 23 billion ($25.7 billion). Foreign direct investment from Germany ceased at the beginning of 2014 according to Russian Central Bank data (Kuchma 2015).

Gazprom, the Russian major global energy company wanted to expand its business by investing in Europe. In 2015, Gazprom Germania announced that it would sell its shares of *Verbundnetz Gas AG (VNG)*, Leipzig, Germany. The 10.52% stake was bought by *EWE AG*, another German energy company (EWE 2015). The decision was based on the geopolitical tensions, a decline in the oil price as well as an increase in the regulation of energy supply networks by the EU. Many Western companies felt the effect of the sanctions. The sales of US automaker *Ford Motor Co.* dropped 40% in Russia (Johnston 2015). Many major retailers suspended sales in Russia altogether after the ruble declined in value. In December 2014, *Apple* halted online sales for its products, only to return to the market in February 2015 with a 35% increase in the ruble-denominated price of the iPhone 5 (Higgins 2014).

Pakistan, Belarus, Serbia, and Chile have benefitted from the counter-sanctions. Russian imports from Pakistan increased by 1.4% in May 2015 compared to the year before, while Serbia shipped 66% and Belarus 53.5% more products to Russia during the same period (Deryabina 2015). In 2014, Brazilian pork exports to Russia, Belarus, and Kazakhstan increased 44.2%. Brazil became the largest supplier of beef to the former Soviet Union, though Russia limited imports of beef from three Brazilian companies due to an "excessive level of antibiotics, *E.coli*, and listeria" found in the meat. It is said that Russia is "making positive steps towards self-sufficiency" in the supply of meat (Vorotnikov 2015).

As geopolitical impacts are felt, business nevertheless goes on. *Siemens* employs 3100 workers in Russia. The Russian business generated EUR 2.2 billion ($2.95 billion) in sales for Siemens in 2013, from projects in the transport, healthcare, and energy sectors. For example, Siemens has a joint venture with Russian Railways to supply 675 freight locomotives by 2020 (Campbell and Webb 2014). In 2014, *Siemens* registered a decline in locomotive orders by one-third. Nevertheless, Siemens does not intend to stop its business activities in Russia altogether; neither does German farm-equipment maker *Claas*, which invested EUR 10 million ($13 million) to add a second factory at their site in Krasnodar and more than double capacity to 2500 vehicles a year (Mangasarian 2015).

DAF puts high importance on the Russian market; parts are needed even though sanctions are in place. *Rhenus*, a major German-based supply chain service specialist, supports DAF in Russia and has launched a simplified customs clearance procedure in conjunction with important representatives of customers and in close cooperation with the Russian customs authorities. "The aim is to continue simplifying and accelerating particular measures such as customs inspections or fixing the customs value of goods within the procedure for the so-called model participants in foreign trade," Rhenus states (2015).

As German exports to Russia dropped by 33% (2015), the German transport industry has been impacted. In the beginning of 2015, the industry estimates to have suffered an up to 40% decline on Russia-bound routes (VerkehrsRundschau 2015). Transport and logistics companies had to face a devaluation and fluctuation of the ruble. Polish carriers were affected too. On average, 600 trucks from Poland enter Russia daily. Since 2015, Polish truck operators require cargo permits in order to move goods into Russia. This caused companies to haul cargo through Belarus or go by ferry through the Baltic States. Due to the significant pressure on the Polish-Russian supply chain in terms of delays, time, and cost, Moscow and Warsaw agreed to each issue the other 20,000 new cargo permits (VerkehrsRundschau 2016).

Beyond Russia, the USA has imposed sanctions on various countries including North Korea, Sudan, and Syria due to their designation by the US State Department as state sponsors of terrorism, human rights abusers, or nuclear proliferators (US Department of Treasury 2015). Restrictive trade sanctions are a way for nations to protect their own business community or specific domestic industries. Trade sanctions are also a strong geopolitical instrument to influence the international or national activities of certain countries. Sanctions are weapons of the modern world (Fig. 1.1).

Sanctions can hit business hard, sometimes even without reason. Take *DF Deutsche Forfait AG*, a German trade finance specialist. In February 2015, the company was placed on the sanction list by the US Office of Foreign Assets Control (OFAC) because of alleged business ties with Iran. The allegations could not be verified and the company was eventually taken of the sanction list. However, the company claimed that this had cost it 9 million euros (approx. $10 million), including 1.5 million euros ($1.68 million) in legal and accountancy fees as well as other costs associated with the OFAC listing. For smaller businesses, such a sequence of events could lead to bankruptcy (DF Deutsche Forfait AG 2014).

Trade sanctions damage a country's economic growth, distort societal progress, and generate uncertainty regarding a nation's future. Extricating themselves from such conditions should be a top priority for affected countries. Iran serves as an

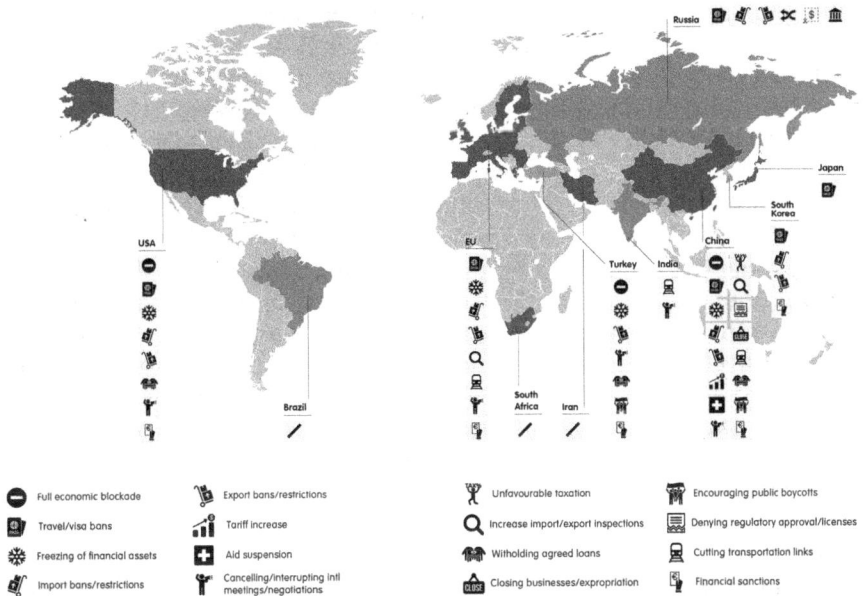

Fig. 1.1 Economic sanctions around the globe. Taken from "The Connectivity Wars." Leonard, Mark. European Council on Foreign Relations (ECFR), January 2016

example: the Islamic republic was subject to various sanctions from 1996 onward. Since 2006, Iran's nuclear program came under scrutiny; in 2012 the EU tightened sanctions by joining the US oil embargo, which was imposed in 1979. The impact can be understood when looking at Iran's GDP, which reached an all-time high of $592 billion in 2011. Following 3 years of sanctions, Iran's GDP stood at just $425 billion by 2014 (Trading Economics 2016).

In January 2016, the International Atomic Energy Agency (IAEA) announced that Iran had passed inspections as required by a nuclear energy agreement entered into by the UN, the EU, the USA, and the Islamic republic. Almost immediately, supply chain service suppliers, from freight forwarders to cargo carriers, traders, and manufacturers, set the wheels of trade back in motion. *Lufthansa Cargo* increased the capacity on flights from Germany to Iran, and freight forwarding company *DB Schenker* resumed engagements with automotive manufacturers. The Iranian government ordered 118 new *Airbus* aircrafts and made agreements for the training of new pilots, airport operators, and air traffic controllers (Woods 2016).

1.2 A Short History of Global Trade

To search for the beginnings of global trade and cross-border supply chains, one must go far back in time. Indeed, this interconnectedness is not a recent phenomenon. Early in history, advances in basic transport technology allowed tribes to establish trade with other tribes, cities with other cities, and regions with other regions. One prominent example is the ancient Silk Road, which was not an actual road but a network of trade routes. First mentioned in 430 BC by Greek historian Herodotus, the 6400 mile-long system was heavily used between 115 BC and the thirteenth century. The Silk Road gets its name from the lucrative trade in Chinese silk carried out along its length by sea and land. Three major routes connected China and Asia with Egypt, Persia, Turkey, Japan, Somalia, and Russia. Routes across the continent started to be less used following the end of Mongol reign. The development of new markets in Southern Asia, high taxation by the Arabs, and the emergence of more modern sedentary civilizations (equipped with gunpowder) played a part in the decline of the Silk Road. The Silk Road network of trade links effectively ceased functioning as a transcontinental transport route in 1453.

Venice, founded in the 400s CE by Roman refugees fleeing the conquest of Attila the Hun, grew into an important trading center by around 810 CE. Strategically located at the far end of the Adriatic Sea, Venice was attractive due to its proximity to several middle European markets. Another economic trade network was the Hanseatic League. The "Hanse" was a commercial and defensive confederation of merchant guilds and their towns, originating in Germany. It was created to protect the economic interests of its members along their trading routes. It was founded in the city of Lubeck at the German Baltic Sea in 1159 and dominated the trade in North Eastern and North Western Europe between the thirteenth and fifteenth centuries. The Hanse also established *Kontors* – literally, offices – along the coasts of the Baltic and North Sea, including Bruges (Belgium), London (England), Bergen (Norway), and Novgorod (Russia). These trading posts offered safe quarters for merchants, while simultaneously expanding the trading network. The port markets also connected the hinterland to the trade routes, which supplied large quantities of timber, wax, amber, resins, and furs, along with rye and wheat. During its zenith, up to 200 cities were part of the *Kontors* network. The Hanseatic League brought both commerce and industry to northern Germany. Because of this economic success, the Hanseatic League grew to be very influential in the political sector as well. Its decline started in the wake of economic crises of the late fifteenth century and with the emergence of territorial nation-states, Dutch merchants, and new overland trade routes.

In Great Britain, an important group of merchants emerged, which was highly influential in regulating the global cotton trade. In 1841, the Liverpool Cotton Brokers' Association was formed, consisting of traders who bought raw cotton from various cotton-growing regions around the world. In 1882, the newly founded Liverpool Cotton Association included merchants and spinners as well as brokers (National Museums of Liverpool 2015). During its height, 80% of Britain's cotton imports came through the port of Liverpool. Here, the raw cotton was processed and resold, bringing wealth to Liverpool and the county of Lancashire.

This became possible due to the First Industrial Revolution, a near global, cotton-based supply chain, made possible by significant improvements in communication and transportation technologies such as transatlantic mail ships and steamboats, as well the telegraph and eventually the telephone. This meant that a broker could actually wire ahead and sell any cotton in transit, well before it even arrived in port. Liverpool became the center of the international cotton trade, connecting raw cotton producers in North and South America, Africa, and Asia with manufacturers and consumers of finished goods. After the World War II, the Lancashire cotton industry went into decline. This was partly the result of a lack of investment in new technology and partly due to manufacturing moving to countries where labor was cheaper. Over the years, cotton processing has continually shifted closer to the places where the crop is cultivated.

In the nineteenth century, trade in Europe changed significantly. European nations expanded their supremacy. From 1800 to 1900, imperialism and colonization lead to an average expansion of each nation's territorial influence to and area ten times its own landmass. It was during this time that European economies changed from being mostly related to agricultural production to being increasingly powered by industrial production. This lead to a growing demand for commodities and resulted in greater trade among nation-states and their respective industries. From 1790 to 1913, the total value of global trade increased 50-fold. Europe imported raw cotton, wheat, meat, and ore from the USA, Asia, and Australia while in turn exporting clothing, textiles, machinery, and train tracks.

Prior to World War I, the largest players in global trade were Great Britain, Germany, and France, accounting for three-quarters of all imports and exports. As a result, European living standards improved significantly. For the first time, products such as sugar, tea, and cocoa became widely available. On the other hand, workers started to become more dependent on developments in other countries and the linkage in international trade.

The World Wars interrupted this development. International trade remained at consistent levels from 1914 onward until the global economic crisis (1929–1932) during which exports and imports declined substantially. During the global economic crisis, high duties and import quotas obstructed global trade and led to price deflation with the prices of commodities declining by an average of 75% and industrial products by 25%.

After World War II, an era of more organized world economy began, led by the USA. The GATT (General Agreement on Tariffs and Trade; 1947–1993) set a framework that eliminated quantitative restrictions and duties. With the introduction of a landmark system for monetary and exchange rate management in 1944, the Bretton Woods Agreement, the United Nations developed rules for commercial and financial relations between the USA, Canada, Western Europe, Australasia, and Japan with the US dollar as the basis. The Bretton Woods system dissolved between 1968 and 1973.

The oil crises of 1973 and 1979, as well as systemic global price inflation, proved to be disruptors to this new global economic order. The oil crisis began with OPEC's decision on October 17, 1973, to reduce production volumes by 5%, resulting in a 70% increase in the price per barrel, from $3 to more than $5.[3]

---

[3]As a unit of measurement, a "barrel" is equivalent to 159 l.

Throughout the course of 1974, the price per barrel of oil increased even further reaching as much as $12. This development resulted in recessions in the developed world with severe global economic effects.

Things turned around again for the global economy in 1980, with an upturn in globalization. The introduction of new technologies was one important reason: micro-electronics and informatics lead to the development of computer and network technologies, which became the prerequisites for global trade. Data on international markets was now available all over the world, instantly, at the push of a button. In various sectors, costs began to decline. Telephone, fax, email, and the Internet eliminated the requirement to produce at a single location. This moment marked the beginning of a wave of offshoring and outsourcing.

Value chains became increasingly segmented as the result of instant communication and high-speed transportation, as well as the possibility to coordinate various steps in the supply and value chain at different locations. According to the Institute of German Economy (Institut der Deutschen Wirtschaft; DIW), globalization was mainly pushed by manufacturers. Thus, the export of intermediate goods increased by 182% between 2000 and 2011, while the export of finished products increased by 138% (Institut der deutschen Wirtschaft 2015). This trend was connected directly to another ongoing economic mega-effect of globalization, the rise of the "emerging markets," in particular Brazil, Russia, India, and China, collectively referred to as the "BRIC" countries.

A pertinent example of this rise can be found in the global cotton industry: the USA was one of the largest cotton growers in the world and an important trading partner of the Liverpool Cotton Association. In 1860, American cotton production accounted for more than 4,000,000 bales (a bale is a compressed bundle of cotton weighing between 400 and 500 pounds) (Dattel 2015). Today, China and India are by far the leading producers of cotton. In 2014/2015 each country produced about 6,500,000 metric tons of cotton, followed by the USA, at 3,500,000 metric tons, in addition to a number of smaller cotton-producing markets, such as Pakistan, Brazil, Uzbekistan, Turkey, and Australia, which contribute to the total global supply (Statista 2015).

Every period of globalization has led to increases in global economic activity which has resulted in improved living standards and wealth generation for people all over the world after an initial period of adjustment to mitigate unwanted consequences such as job loss/change and income inequality. On the other hand, disengagement from global trade and production networks has generally shown negative economic impact, one clear example being Cuba. Empirical data seems to reflect that a major prerequisite for positive economic development is the fluid exchange of knowledge, goods, capital, talent, and culture.

Currently, the historic Silk Road is experiencing something of a renaissance. In September 2013, Xi Jinping, President of the People's Republic of China, proposed the Belt and Road (BR) initiative to China's neighbors, with the aim of adding over $2.5 trillion in annual trade over the next 10 years, with the nations located along the proposed route. The BR initiative, also called the "Global Economic Strategy," has two main components: first, the Silk Road Economic Belt, or SREB, encompassing the Eurasian landmass and, second, the twenty-first Century Maritime Silk Road, which encompasses maritime routes connecting China, ASEAN, and Eurasian countries to Europe. Major investments have been made along the Maritime Silk Road to improve port access and connectivity. This land bridge is intended to provide alternatives to ocean and air transport between China and Europe, reducing the transit time of goods from around 60 days by sea, to 14 days by land (Lehmacher and Padilla-Taylor 2015).

The BR needs to overcome technical and regulatory challenges: heterogeneous customs clearance procedures and at least two changes of rail gauge. China and Europe work on the basis of the normal standard of 1435 mm, while Belarus, Russia, Mongolia, and Kazakhstan use the broad gauge of 1520 mm. The BE requires the revamping and upgrading of existing infrastructure, as well as efficient and simplified customs procedures and high-performance digital infrastructure and processes to match. To achieve this, roads, bridges, railways, airports, oil and gas pipelines, power grids, maritime links, and Internet networks will require major infrastructure funding.

Involving an area that generates 55% of global GDP and has 70% of the global population and 75% of the known energy reserves, the future BR web will have a major impact on all the dimensions of trade and the economic development of the participating economies. The BR promises to transform the geopolitical landscape by creating incentives for the establishment of new trade lanes, reviving local industries, and generally creating new economic opportunities along the route.

This new Silk Road may have actually reached a tipping point, in terms of likely realization, as a result of China's current economic, financial, and technical ascendance. With the BR, China commits, in its own and the common interest, to forge a community of common destiny where peace, cooperation, openness, inclusiveness, and benefit sharing will improve the performance of existing supply chains and trade, while also increasing food security in the region. Investments are underway. China is planning to build a high-speed railway system that will stretch over 16,000 km once completed. As of January 2014, construction of sections connecting China with Vietnam, China with Myanmar, and Laos with Vietnam are ongoing. In December 2015, construction began on a 427 km segment of railway that will link the Laotian capital, Vientiane, with the Chinese border. The $6.04 billion Laos-China railway construction joint venture is expected to be completed in 4–5 years (Ghosh 2016). The Chinese government is also making significant investments to improve infrastructure in Africa. The goal is to link China and Africa by building a railway system to Urumqi in Western China. In 2011, German-based DB Schenker and the Chongqing Holding Group, plus the state-owned railways of China, Kazakhstan, and Russia, formed a joint venture. One of the joint venture's customers is US computer company Hewlett-Packard. The company has shipped more than four million notebook computers from factories in China to Europe by way of the China-Europe rail corridor (Nurshayeva 2013).

Ultimately, North America might be connected by rail to China as well. The proposed rail line would lead from Northern China, across the Bering Strait, to Alaska and into Canada and would include a 200 km-long undersea tunnel. Since the high-speed trains could reach speeds of 350 km an hour (217 mph), China would be linked to the East Coast of the USA in 2 days (The Siberian Times 2014a, b). In addition, there is the proposed China-Pakistan Economic Corridor (CPEC). This network would consist of about 3000 km (1864 miles) of highways, railways, and pipelines connecting China's Xinjiang province to the rest of the world through Pakistan's Gwadar Port on the Indian Ocean. This project, worth approximately $45 billion, aims at improving the trade routes between South and Central Asia (The Pakistan Times 2016).

The realization of this new Silk Road calls for efficient logistics services along a vast economic corridor. Europe and the USA are home to major logistics and transportation companies offering global operations, footprints, and services. In China, the supply chain service market is widely fragmented, with quality of the services varying from one city to the next. Experts estimate that approximately 700,000 logistics firms offer their services in China, many of them being one-man operations. The 20 largest logistics players in China hold a combined national market share of just 2% (The Economist 2014).

As few goods merit the speed, and high cost, of air transportation, most cargo is shipped by sea. Between Asia and Europe, the ocean voyage takes around 30 days, not including inland transport from and to the ports. Rail offers a value proposition somewhere between air and ocean freight. The southern rail route between China and Russia, through Kazakhstan (6213 miles), takes 12 days, while a northern rail route through Russia (8077 miles) takes 16 days, either. Can the Hyperloop ultimately provide the quickest transportation option, at affordable cost?

In June 2016, Anthony Cuthbertson wrote in *Newsweek* that "Vladimir Putin imagines a Hyperloop Silk Road," where products might soon travel over land from Chongqing in China to Stuttgart in Germany in less than a day. In 2013, inventor and tech entrepreneur Elon Musk envisioned a vacuum tube to move passengers and cargo in 35 min from San Francisco to Los Angeles (Grothaus 2016). Two companies – *Hyperloop One*, former Hyperloop Technologies, and *Hyperloop Transportation Technologies (HTT)* – are exploring Musk's idea and planning to build the first vacuum tube for transport by 2020 (Lipton 2016). According to Hyperloop One Chief Executive Officer Rob Lloyd, feasibility studies have been launched not only in Russia but also in the United Arab Emirates, Finland, the UK, and the USA. Hyperloop One has joined forces with the city of Moscow and Summa Group, a local investment and construction conglomerate, to bring the vacuum tube concept to Russia. Much as oil and gas resources travel through pipelines, cargo and passengers may soon travel through vacuum tubes at speeds up to 760 mph. Summa Group is well positioned to support the Hyperloop One project; the company owns a logistics business and has experience constructing oil pipelines, a competence that could be highly valuable in building the first transonic tube.

Hyperloop construction costs are important to consider. Building and operating Hyperloop are estimated to be in a similar range as building and operating high-speed rail infrastructure and trains. The construction cost for high-speed rail lines, such as the Haikou-Sanya line in China and Madrid-Albacete line in Spain, comes in at about \$15 million per mile (Railway Technical Web Pages 2016). Elon Musk suggested that a San Francisco-Los Angeles Hyperloop would cost \$11.5 million per mile (Nicas 2016). Hyperloop One estimates cost to build the vacuum tube to be between \$5 million and \$20 million per mile. However, the cost might be much higher if, instead of pressurized air generated by compressors as proposed by Elon Musk, the more expensive maglev technology is installed, using powerful electromagnets to lift the train cars. The 19 mile maglev line in Shanghai has cost \$63 million per mile, and a proposed 178 mile track in Japan is expected to cost roughly \$250 million per mile. In early 2016, HTT announced an exclusive deal to license passive magnetic levitation technology, which would eliminate the need for power stations along the track and therefore lower operating costs. HTT had also filed permits in Kings County, California, to build a 5 mile test track (Hawkins 2016). Hyperloop One tested the propulsion mechanism on May 11, 2016, in Nevada. Although the test result can be considered a first proof of concept, the outcome cannot yet be called a Wright Brothers' "Kitty Hawk" moment, as the demonstration focused only on one part of the complex solution (Farrington 2016).

The race is on, though. The transonic Silk Road project would require sizable political and financial support from China. Attracting China's interest in Hyperloop will be instrumental to building a transonic land bridge between China and Europe. Initial interest seems to exist. In January 2016, *Bloomberg* reported that, according to people familiar with the matter, CRRC Corp., China's largest maker of railway equipment, was in talks for a potential investment in Hyperloop One (Bloomberg 2016). Depending on the developments, the Belt and Road initiative might soon connect Asia and Europe overnight by the Hyperloop, potentially necessitating a name change from "Belt and Road" to "Belt and Tube."

1.3   Trade in the Interconnected World

The world is more connected than ever, by way of digital flows. According to the McKinsey Global Institute, cross-border digital traffic volumes have grown 45 times between 2005 and 2015 and are expected to grow by a further factor of 9 by 2020. The institute estimates that 900 million people around the world have international connections on social media, while about 360 million take part in cross-border e-commerce (Manyika et al. 2016).

Global logistics company *Deutsche Post DHL* analyzed data flows and the resulting impact on prosperity. Published in the DHL Global Connectedness Index 2014, the analysis covers 140 countries, which encompass 99% of the world's GDP and 95% of its population. The study focuses on 12 types of trade, capital, information, and people flows (or stocks accumulated from past flows). The study found that following the global financial crisis (GFC) in 2007/2008, the intensity of international interactions and global capital flows reached precrisis level by 2013.

Europe is the top-ranked region in terms of overall global connectedness and leads on the trade and people pillars. North America is the most connected region on the capital and information pillars. The countries with the largest increases in their global connectedness scores from 2011 to 2013 are Burundi, Mozambique, Jamaica, Madagascar, and Suriname. Economic centers are shifting, as more and more nations participate in the global market, gradually creating a more balanced and inclusive world.

1.3.1 Rising Complexities

Supply chains have become highly complex. For a long time, the main function of the supply chain was the movement of raw materials, parts, and finished products to and from major industrial nations and consumers around the world. The task was carried out by a limited number of participants along the chain, upstream by a few suppliers and downstream to a few major wholesalers.

Supplies are ordered from more suppliers than ever before, with unprecedented choice, cost benefit, and risk distribution. This brings new challenges in terms of supply chain planning, management, and orchestration. Downstream, wholesalers have been replaced by retailers, which are increasingly replaced by direct consumers. Compliance with laws and regulations require all suppliers are identified, mapped, audited, and properly monitored – from tier 1 to tier n – in order to better understand dependencies and potential risks. And risks, real or perceived, are increasing. According to the Chartered Institute of Procurement & Supply (CIPS) Risk Index, global supply chain risks increased in the fourth quarter of 2015, "resuming a worsening trend in global operational risk" (CIP 2016) (Fig. 1.2).

Downstream at the customer interface, the digital revolution is significantly enhancing the interaction, exerting huge pressure on traditional modes of operating and business models. New combinations and innovative products are required. Electronic and mobile commerce are dramatically changing the way consumers and companies buy. Today, online and offline go hand in hand; e-commerce and m-commerce are joined by online-to-offline. In the omni-channel world, consumers expect that anything is available, anywhere and anytime. Deliveries at home, the office, service points, lockers, and boxes are the new normal. In high-density urban areas, deliveries are expected within minutes – 24/7. Many e-commerce companies offer free returns. Consequently, high-performance and convenient reverse logistics solutions are in high demand.

In parallel, the range and variety of products is increasing. In SCM World's Value Chain 2020 survey, 90% of organizations feel their customers "strongly value" or "value" individualized products (Manenti 2016). Parts of a car such as the steering wheel are being customized: customers can choose a leather cover, internal heating, aluminum or steel elements, and various functions or controls. For managers in procurement/sourcing, marketing and sales, logistics, and transport, this raises many new questions: How can companies sell and deliver products anytime to customers anywhere in the world? What kind of supply and value chains are needed? Are retailers and e-commerce companies better off establishing their

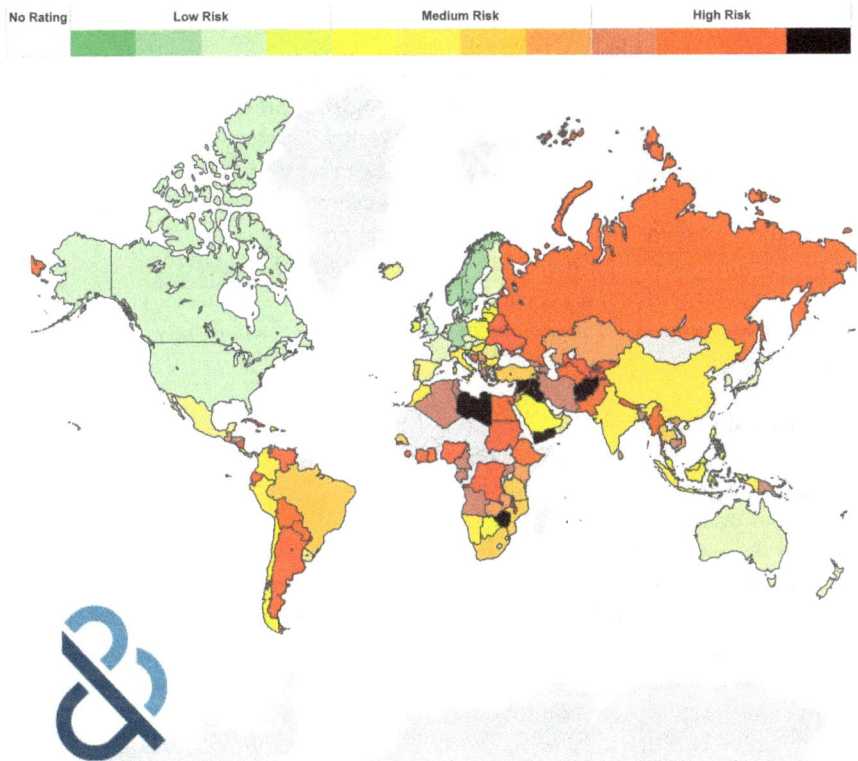

Fig. 1.2 Global supply chain risk in Q4 2015. Source: Dun & Bradstreet, CIPS Risk Index/ Quarterly Report Q1 2016

own delivery and return capabilities or even controlling the entire chain end to end? How can modern ICT and digitization help?

1.3.2   Digital Revolution

Nearly 90% of respondents to a 2015 global survey of managers and executives conducted by MIT Sloan Management Review and Deloitte anticipated that their industries would be disrupted by digital trends to a *great or moderate extent*, but only 44% say their organizations are *adequately preparing* (Kane et al. 2016). Information and communication technology (ICT) is a key enabler of the modern global supply and value chain. Today, ICT provides information visibility, instantly and across long distances, making highly efficient and decentralized processing possible. Information visibility and analytics provide further levels of insight and greater efficiency. Automated customs clearance systems, such as the Automatic Rate and Local Customs Clearance System (ATLAS) in Germany, the integrated Tariff of the European Union (TARIC), or the US Automated Export System

(AES), are examples of digital technologies improving the efficiency of exchange between nations and markets.

E-logistics is the new business model. The IBM Watson Research Center defines e-logistics as "the mechanism of automating logistics processes and providing an integrated, end-to-end fulfillment and supply chain management services to the players of logistics processes. Those logistics processes that are automated by e-logistics provide supply chain visibility and can be part of existing e-commerce or workflow systems in an enterprise."[4] Through e-logistics, the movement of goods becomes faster and more cost efficient, and the logistics value chain more resilient: potential disruptions can be identified early. E-logistics is particularly important since global supply chains are becoming increasingly complex and interconnected.

Supply Chain Visibility

Performance and visibility in the supply chain are closely connected: the higher the visibility, the greater the possibility of optimizing efficiency and avoiding potential disruptions. Without the knowledge of where certain orders are or what kind of products is produced in which part of the world, chief supply chain officers (CSCOs) cannot act to mitigate potential disruption, for example, in the event of political unrest or natural disaster. To plan ahead and prepare for various scenarios, CSCOs rely on historic data and timely information. Visibility also promotes greater trust from customers and consumers, because tracing a product's source of origin, material mix, carbon footprint, and reusability, all matter.

Among the benefits of increased visibility is the possibility of enabling vendor-managed inventory (VMI). Visibility can also reveal important information on product quality by monitoring which products and parts are returned and for what reason. Visibility can help to reduce disruptions. However, although 42% of all supply chain disruptions occur beyond the direct control of supply chain partners, 75% of all companies have no insight into the sublevel structures of their supply chain, according to the study "BCI Supply Chain Resilience 2015."

David Simchi-Levi, a professor of engineering at the Massachusetts Institute of Technology, together with William Schmidt of Cornell University and Yehua Wei of Duke University, has illustrated what the potential impact of unexpected disruption at one supplier's site could have on a company's operations. Their methodology, which was implemented by Ford Motor Company, quantifies the financial and operational impact of supplier disruption risk. According to David Simchi-Levi, a central feature of the model is time to recovery (TTR): the time it would take for a supplier facility, a distribution center, or a transportation hub to be restored to full functionality after a disruption. By combining suppliers' TTR information with the details of Ford's supply chain, the method identifies the risk exposure associated with a disruption in each site of Ford's network. The research found that suppliers

---

[4]For a more extensive definition, refer to http://university-essays.tripod.com/e_logistics_distribu tion_and_supply_chain_management.html

tend to be optimistic about their TTR since they know that a long TTR will not be accepted by the manufacturer. This discovery led to the implementation of a new metric called "time to survive" (TTS). It is the maximum duration supply that is ensured after a node disruption. If the TTS is less than the TTR, problems are likely to arise. A TTS greater than a TTR suggests a more robust supply chain (Simchi-Levi 2015). Ford Motor Company is using this method to mitigate supply chain risk exposure on an ongoing basis. Once the inventory falls below an identified value, an alarm is triggered and order management is set in motion. This method allows Ford to define supply assurance strategies and measures with its thousands of suppliers and create a more robust supply chain.

Of over 2660 supply chain professionals surveyed by Deloitte, 31% suffered from supply chain fraud between January 2015 and January 2016. Almost 20% of the same professionals declared that they have not started using analytics to mitigate fraud and waste and abuse risks in their supply chain. Only 8.6% responded that they are using analytics to manage third-party relationships (Deloitte 2016). A whitepaper entitled *Managing supply chain risk in a flat world* points out that, in today's fast moving world, changes occur all the time and risk analysis needs to be conducted regularly. Systems offering a top-level view of all the information at a glance are of immense help (Jamison and Kain 2016).

Some experts point out, however, that one has not only to consider major disruptions, such as natural disasters or geopolitical conflicts, but also "micro-risks." These could range from computer viruses, and even seemingly positive events, such as technological advancement, which can pose a risk to those companies that cannot keep up with the pace. As supply chain managers mostly deal with occurrences likely to happen, some events will leave companies unprepared, without the appropriate precautions and contingency plans in place. Companies need to establish intelligent solutions that deliver concrete options for action. The best way of mitigating micro-risks, though, is to prevent them from happening in the first place. Here, resilience is the key once more (Keppner 2015).

Transparency contributes to an atmosphere of greater trust and helps establish a common ground for collaboration. It also helps with respect to compliance and risk management. Compliance risks are reduced by creating more visibility along the chain. If capital flows can be fully traced, corruption is less likely to occur. The process of awarding contracts to tier-one to tier-n suppliers, for one, needs to become more transparent. Visibility allows supply chain managers to assess risks and violations of law or the code of conduct more easily. In a recent market analysis, the advisory group ARC notes that supply chain visibility and collaboration will be inevitable in the future and will be the fastest-growing segment across a range of supply chain solutions. This is due to the growing number of B2B networks engaged in supply chain planning (SCP). ARC indicates that fast growth will persist because major software companies, such as *SAP* and *Oracle*, are developing solutions for this promising market segment (Banker 2016).

The Internet of Things

The Internet of Things (IoT) will also create a new world of opportunities and business models for the supply chain. The Internet of Things, the concept of billions of connected devices providing data which can be analyzed and used for planning and control purposes, has taken the Internet to the next level. The IoT will provide an unprecedented degree of supply chain visibility to manufacturers, consumers, and business customers, not only on the flow and status of deliveries but also the condition of shipments and merchandise.

Consider the following real-world example (Lehmacher 2016): During transit of a thermal pallet, Henry Schein Inc., one the world's largest providers of healthcare products and services, was experiencing significant delays in customs clearance at a location in the Middle East. The important shipment of perishable vaccines and antibiotics was at risk of becoming medical waste. Data collected by sensors in the shipment alerted the company that the packaging was reaching thermal limits. Through close coordination with its transportation partner, the shipment was quickly reconditioned to preserve the viability of the vaccines and antibiotics and ensure its safe delivery. Smart packaging prevented this shipment from spoiling in transit, which would have caused a significant shortage of important medical supplies in the affected area.

Supply chain visibility (SCV) is the basis of entirely new business models and an enabler of the circular economy. SCV allows the traceability of component parts from material suppliers and manufacturers to wholesalers, retailers, and the final customer, enabling visibility even beyond the first use cycle and creating new opportunities for product reuse, remanufacturing, and recycling. The goal is to optimize all processes along the chain, minimize risk, and preserve or increase the original value throughout all use cycles. Out-of-use parts and products can be injected into subsequent or new supply and value chains. SCV increases the available data for managing the flow and use of goods, as well as the analysis to allow better recommendations, control, and strategies to improve products, services, and the supply chain. Data is made available to all relevant and authorized players along the supply chain, including customers and government agencies. Technology can establish visibility to help brands monitor the activities and condition of the goods handled, carried, and transformed along the supply and value chain. One example of high visibility and consumer transparency is *Fairphone*, a company that allows any visitor to their website to view a comprehensive list of suppliers, locate sources of materials, and track the flow of goods.

Merely monitoring upstream and downstream within the supply chain is not sufficient. Scenarios need to be developed, and contingency plans to be at hand and be able to act quickly as soon as challenges become evident. Big data and predictive analytics can help to understand the probability of potentially disruptive events. Visibility can be established and extended to areas and developments far beyond the operational supply and value chain, to the entire ecosystem. Geopolitical developments, social upheaval, strikes, or natural disasters can be constantly monitored and analyzed to allow for adjustments in the operating plan or the launch of alternative plans. Visibility provides the basis for modern supply chain management. ICT delivers platforms for company-wide collaboration and information exchange, helping companies to design, control, and deal with challenges.

We have entered the era of platformization. Let's take the example of procurement or sourcing; companies can post their needs; suppliers, traders, and brokers can submit proposals. The same platform can be linked to various internal and external systems. Information about the status of the order process, weather reports and contingency plans, can be made available. *Uber*, the mobile transport service provider and platform, enables the on-demand movement of passengers, as well as goods-delivery service. In response, traditional taxi companies have been increasingly leveraging mobile apps to remain competitive.

The IoT and platforms can take control over entire supply chains and manage our stocks and the replenishment at home and at the workplace. Running out of coffee or razors can be frustrating and throwing away expired medicines and food too. Managing inventory of all kinds even in the smallest of households is an art perfected by few. Monthly subscriptions for deliveries on a regular basis offer some support. With five billion connected "things" in 2015, however, there should be better solutions to make empty stocks and out-of-date products a thing of the past (Gartner Inc. 2014). Household appliances are becoming increasingly connected and equipped with sensors to provide a broad range of information to users and other parties. For example, a prototype of a smart, reusable milk cap that uses sensors to detect when milk starts to go bad was recently created by researchers at the University of California-Berkeley (Peters 2015).

Sensors and smart appliances can also help us replenish our groceries. One important step towards auto-replenishment is *Amazon* Dash Button, launched on April 1, 2015. Once pushed, the "Button," attached to appliances around the home, processes automatic refill orders and delivery requests. Each time the button is pressed, an alert is sent to a mobile phone connected to the account, and, once the user has approved the order, detergents, coffee filters, and other essential articles are delivered to their home, some even within minutes.

More importantly, alongside the Button, the Dash Direct Replenishing System was launched. This service provides an Application Program Interface (API), a specification which allows smart devices to communicate with each other (Sareen 2015). The API enables manufacturers to directly integrate the Dash system into appliances, bypassing the Button. Without any human intervention, the espresso machine might soon know when and where to place the order when the stock is low. No more frustrations, no more time wasted writing the purchasing list for our day-to-day products.

Replenishment services are not only for homes either. Any business or organization can benefit from this development, and companies can design new offerings. This will drive additional business too. Xerox, for example, has recognized that the most common reason people can't print is that the printer has run out of toner. Therefore, the company has launched an auto supplies replenishment service for printers (Xerox 2014). And this is probably only the beginning of a massive development of new services driven by connected devices and the capabilities offered by the Internet. What if household appliances booked not only the orders for their refills but also the spare parts in the case of a breakdown and notified the nearest repairman too?

Of course, everything connected to the Internet can be hacked. After two hackers were able to take control of a Jeep over the Internet, 1.4 million vehicles needed to be recalled (Associated Press, *The Guardian* 2015). In the hyperconnected world, our security is at stake, both at home and in the workplace. The risk lies not only in the possibility that hackers can take control of devices but also of the data gathered by the devices, whether it be sensitive personal information or the secret research data of a company. The route towards tomorrow's world of increased convenience and efficiency needs to be explored and the relevant precautions be taken, not only by individuals but also by companies, particularly the manufacturers of smart devices.

What will the future of the highly connected world look like? Will we see bright and colorful branded buttons on appliances everywhere in our homes, offices, and other organizations? Probably not. The likely mainstream scenario is the gradual increase of the number of connected "things" controlled by smartphones, tablets, and wearables, acting more and more autonomously. The Apple HomeKit is an indication of the type of framework required to control this world of partially autonomous appliances. The Kit might be one of tomorrow's most stylish housekeepers.

The Digital Consumers

Through the Internet, consumers have easier access to goods and a wider range of information on products, labor and manufacturing conditions, and environmental impacts. Consumers and citizens have a clearer picture of the situations and events along the supply chain. The world has gotten smaller. A generation ago, global events actually seemed to be a world away. Now, political unrests or ecological incidents are transmitted into our living room and our smartphones. Today's consumers can easily imagine how their purchasing and consumption behaviors influence the global economy, society, and environment. With this knowledge comes responsibility. Informed citizens are not passive spectators but co-shapers of global events. Consumers understand that their consumption choices impact the exploitation of raw materials and the climate. Through targeted interventions and with the assistance of the Internet, consumers can influence social situations and economic outcomes at home and abroad.

Leading by example, business should redefine its role in the supply chain and encourage consumers to do the same. Consumers are empowered by information to make more responsible purchasing decisions, which have upstream impacts on manufacturing processes, procurement/sourcing, and ultimately product design. Consumers possess a mighty voice. In the event of deviation from publicly communicated code of conduct, corporate social responsibility (CSR) statements, or the brand promise, consumers can act swiftly and aggressively. Customers can boycott particular products or brands, holding companies to account for socially irresponsible behavior. Customers can leverage various social media platforms to press firms on products, sourcing, service, or operations, instead of waiting for the media or politicians to take up the case. Customers can catalyze media response and follow up on particular issues. In the digital age, information can rapidly go viral. People share or post original content across various platforms on the Internet. This can precipitate actions, such as so-called flame wars, that could actually cause severe economic losses to the brands. There have been numerous boycotts which resulted in a change in company practice.

In 2010, clothing manufacturer *Fruit of the Loom* had to submit to pressure from student boycotts in the USA and UK: In 2009, a group called United Students Against Sweatshops launched a campaign which led a number of US colleges to terminate contracts with Fruit of the Loom. British universities followed suit. The reason was that a factory in Honduras had been closed and workers had been laid off after they had formed a union. Fruit of the Loom gave all 1200 employees their jobs back, awarded them $2.5 million in compensation, and restored all union rights. The campaign was estimated to have cost the company $50 million (The Ethical Consumer 2016). Another example involved *Nike*. In the 1990s the sports company was accused of employing children. Nike's top line took a major hit and the brand reputation suffered. Since then, Nike has reinforced its policies and is positioning the brand as a sustainability leader. Judith Samuelson, executive director of the Aspen Institute's business and society program, told *The Guardian*: "Today, Nike is very proactive. They're aware of the need to be on top of their supply chain" (Watson 2015). *Apple* seems to be an exception to the rule. Though the corporation has been subject to criticism due to labor conditions at its supplier Foxconn, it does not seem to impact Apple's reputation or sales.

Making a statement is high on the agenda of consumers who invest in eco-friendly products, such as *Fairphone*, *Brooks* BioMoGo sneakers, or hybrid vehicles. Fairness and responsible behavior are the purchasing arguments. The new visibility has changed the probabilities. Everything seems to come to light, everything can happen, sooner or later: whether good or bad, right or wrong. Therefore, businesses should develop and adhere to a comprehensive proactive sustainability and communication strategy by integrating traditional and new media. Honesty and transparency should be the new mantra.

3D Printing
While not new, 3D printing has increased in use and popularity. Also referred to as additive manufacturing, the process involves making three-dimensional solid objects by applying layer on layer of a material. The technology offers alternative means of manufacturing and prototyping: a 3D printer can download blueprints off the Internet and make objects and parts, which can be customized.

There are three major types of additive processing, differing only in the way layers are built to create the final object. The most common technologies are stereolithography (SLA), selective laser sintering (SLS), and fused deposition modeling (FDM). According to the Wohlers Report 2015, the worldwide 3D printing industry is expected to grow from $3.07 billion in revenue in 2013 to $12.8 billion by 2018, to exceed $21 billion in worldwide revenue by 2020 (3Dprinting.com 2015; Wohlers Report 2015).

3D printing unlocks new markets, with consumers even becoming producers or so-called prosumers. The production process is no longer limited to a certain location or region. Patents, trademark protection, and legal restriction still need to be respected – an area policymakers and governments will need to tackle. Issues such as intellectual property rights, patents, and documentation, which would include producing blueprints, as well as legal liability issues related to the printing of prescription drugs will have to be clarified.

PricewaterhouseCoopers indicated that two-thirds of more than one hundred industrial manufacturers surveyed are already using 3D printing. Most of them are still experimenting with this emerging technology (28.9%), while some companies are already producing prototypes (24.6%) or are both prototyping and producing (9.6%). However, <1% is actually manufacturing final products or components. The PwC survey highlighted growing demand for additive manufacturing in the following industries: automotive and industrial manufacturing (e.g., spare parts and components, production tooling), aerospace (e.g., lighter component parts), pharmaceuticals and healthcare (e.g., bioprinted live tissues for testing during drug development, custom orthopedic implants, and prosthetics), retail (e.g., custom toys, jewelry, games, home decorations, and other products), and sports (e.g., custom protective gear for better fit and safety) (Earls and Baya 2014).

Forbes pointed out that 3D printing will become "industrial strength" and will represent a competitive advantage as "customization becomes the norm" (Srinivasan and Bassan 2012). Deloitte calls additive manufacturing "a driver of supply chain transformation" for the automotive industry because it will "cut down on overall lead time, thus improving market responsiveness." Furthermore, 3D-printed, lightweight components can lower handling costs, while on-demand and on-location production can lower inventory costs (Giffi et al. 2014).

Retailers are becoming increasingly aware of the opportunities 3D printing has to offer. Major sporting shoe brands, such as *Adidas*, *Nike*, *New Balance*, and *Brooks*, are experimenting with the technology, customizing the fit and the lightness of shoes even further. The market also saw the emergence of innovative start-ups. New York-based *Sols* are already offering customized 3D-printed orthotics. *Feetz* makes 3D-printed shoes based on photos of shoppers' feet (Thumm 2016).

The port of Rotterdam has become home to an Additive Manufacturing Fieldlab. The port will experiment with metal printing, 3D scanning, and 3D design and eventually establish a certification process. "This Fieldlab will provide port-related companies with a collective location to accelerate developments in this area and to work together on applications for the (maritime) industry," Allard Castelein, CEO of the Port of Rotterdam Authority announced in February 2016. The port of Rotterdam established partnerships with various universities and project partners to reach the goal of making the "industry more competitive by means of exploiting the opportunities offered by ICT" (Port of Rotterdam Authority 2016).

Additive manufacturing also opens new markets for after-sales services to optimize and expand a company's service structure. In 2014, *Amazon* launched a 3D printing store, partnering with several 3D printing production companies, including Cincinnati-based *3DLT* and Brooklyn's *Mixee Labs*, to print figurines, jewelry, home decor, toys, and tech accessories (O'Toole 2014). With this approach, Amazon is following *Shapeways*, a pioneer in the 3D printing market-place and service company, which has printed and sold more than one million user-created objects (Raven 2012). 3D printing allows consumers to produce on-demand and at home. Customers can source parts more responsibly and produce less scrap, since production is launched when needed rather than produced in advance and kept in stock for potential sales which might never happen. Amazon even wants to take 3D printing technology on the road. In 2015, Amazon Technologies, Inc., was granted a patent which outlines a method of 3D printing on-demand within mobile manufacturing hubs. The idea "Ordered goods would be printed inside a delivery van while on the way to the customer" (U.S. Patent & Trademark Office 2015) will make the supply chain even leaner, and the warehouse turns into a moving production unit.

Through 3D printing, brands can enhance their after-sales service by providing spare parts without time limit, prolonging the object's life cycle and causing less waste, while simultaneously clearing their warehouse space and thus reducing fixed costs for real estate. A wide range of firms can benefit from these new market opportunities. Customizing products or manufacturing parts on-demand, which would make stock-keeping redundant, are potential areas of interest for logistics companies. Consumers, business customers, manufacturers, and distributors can benefit from shorter production cycles; manufacturers can also reduce expenditures on tools through 3D printing.

Cloud Computing

The proliferation of cloud-based enterprise applications and the relative affordability of massive online storage capacity mean that powerful cloud comput-ing services are no longer the exclusive domain of large enterprises. Almost 80% of U.S small businesses will be adapted to cloud computing by 2020, according to "Small Business Success in the Cloud," a 2014 report from Emergent Research and Intuit Inc.

Of course, large companies also benefit from cloud computing. *Pfizer*, the American pharmaceutical company, uses a cloud-based platform to connect to all of their freight forwarders, carriers, and other logistics service providers. Service providers are required to send their data to the cloud, so Pfizer can track and trace its shipments worldwide. This virtual supply chain has vastly improved visibility and allows Pfizer to take action the moment disruptions to the supply chain emerge. Since certain vaccines have to be refrigerated while in transit, Pfizer has to carefully control temperature throughout the shipping process. Using its cloud-based logis-tics management platform, Pfizer is able to do just that, as well as apply advanced analytics to enhance supply chain efficiency. Supply chain managers can more easily drive quality improvements and cost reductions in the network. With an integrated, global view of the supply chain, Pfizer's global transport team can easily spot opportunities for consolidating shipments to save cost and prevent supply chain disruptions (Supply Chain@MIT 2013).

Cloud-based platforms can provide enterprises high levels of visibility into their supply chains and at relatively low cost. There are many different options available to enterprises when it comes to cloud services, such as infrastructure as a service (IaaS), platform as a service (PaaS), or hybrid solutions, combining public and private cloud offerings. Cloud computing decouples enterprises from the need for in-house software development, maintenance, and IT management processes. Enterprises big and small are usually allowed substantial flexibility in their service contracts, adjusting capacity and scaling down or up as needed. The main challenge to cloud-based platforms is not the storage space or service itself but rather the integration of the cloud-based platform with legacy IT systems. Ultimately, most existing legacy IT systems, such as enterprise resource planning (ERP) systems, warehouse management systems (WMS), transport management system (TMS), and document management systems (DMS), will be moving to the cloud.

Still to be settled is the question of global standardization for cloud environments. In 2014, the International Organization for Standardization (ISO) released two new standards for cloud computing, the ISO/IEC 17788 and ISO/IEC 17789. The standards set rules regarding security issues as well as personal data. The IEEE Standards Association (IEEE-SA), an entity of the Institute of Electrical and Electronics Engineers (IEEE), is also working on cloud computing standards.

Spotlight

The German *Fraunhofer Society for the Promotion of Applied Research* (Fraunhofer-Gesellschaft zur Förderung der angewandten Forschung e. V.), based in Munich, Bavaria, has developed cloud-based solutions for the logistics industry. Two of its institutes, Institute for Material Flow and Logistics (IML) and Institute for Software and Software Engineering (ISST), have developed a concept which translates as the "Logistics Mall." Logistics is the third largest industry in the German economy and employs 2.7 million people (equivalent to about 7% of total German workforce). The Logistics Mall aims to support the movement of key logistics processes, ranging from order management to invoicing, onto a globally accessible cloud-based platform. The Logistics Mall includes four areas: the mall marketplace (MMP), the customer process and service portal (CAF), the logistics process designer (LPD), and the service engineering framework (SEF). Standard processes related to managing inbound goods and commissioning, for example, can be customized and assembled on the cloud to support a unique supply chain.[5]

---

[5]Information on the Logistics Mall can be found at http://www.ccl.fraunhofer.de/en.html

The global supply chain needs cloud development acceptance and global standards. The supply and value chain will benefit enormously from the possibilities the cloud offers.

Big Data

The term big data refers to the aggregate of data from various digital sources as the basis for use and analysis. It is called "big" because the amount of data is too large to be processed with traditional databases and software tools. However, big data does not refer solely to the volume but also to the tools and processes used to handle unstructured data from various different sources.

According to a 2015 survey by the APICS Supply Chain Council, 52% of survey participants felt that gathering data from multiple sources is their biggest challenge. Forty-nine percent reported difficulties working across different data formats, standards, and files. Also about half of the survey participants felt that there was too little relevant data upon which to build reliable forecasts (APICS 2015). But big data can be a very powerful tool.

Amazon, for example, collects customer data such as IP addresses in real time, while also storing information on user preferences and browsing habits to derive and present recommendations. Recommendations, such as "customers who bought/viewed this item also bought/viewed" and "this item is often sold with," are based on big data and the use of complex statistical algorithms. If customers have liked Amazon on Facebook, the retailer also draws data on any FB friends' Amazon purchasing behavior to make more recommendations.

Data analyses can also be used to reduce costs in sales and marketing. The U.S department store *Macy's* uses big data to ensure competitive pricing for 10,000 items across 800 stores. The company takes various factors into account, on a Macy's store-by-store basis, including the prices of local competition, Macy's special offers, and storewide sales at a given location. If a local competitor begins carrying a particular product, Macy's reduces prices for that product in its nearby stores. All in all, around 270 million near-real-time data points are captured and analyzed, completing in 2 h an analysis that, just a few years ago, would have taken 30 h to finish (T-Systems 2016).

Knowingly or unknowingly, consumers and competitors have been in the crosshairs of supply chain and marketing managers for some time already. Anticipating customers shopping behavior helps to reduce cost, mitigate risk, and ensure on-time delivery of products and components along the supply chain. *DHL*, the global logistics company, uses applications for volume and capacity planning. A tool called Parcel Volume Prediction maps the impact of weather patterns and communicable disease pathways on online shopping behavior. The model predicts the impact these patterns might have on the volume of packages delivered over a given time period and allows for more effective capacity planning (DHL Global 2014). DHL also uses a tool called "Resilience 360" which collects data on political and economic conditions in certain regions to predict potential areas of supply chain disruption. This tool helps DHL prepare contingency plans well in advance to avoid down time and additional cost along the supply and value chain.

Manufacturers also use big data to optimize their supply chains. *BMW*, the German car maker, has connected each of its 4500 suppliers and partners to its digital supply chain management tool. This provides BMW the capability to more efficiently plan production output and manage second-tier suppliers. Data across BMW's supply chain is gathered, refined, and analyzed in real time. "The sooner we know about disruptions or delays, the faster we are in measuring their effects on the supply chain and can take countermeasures," Johann Gerl, Head of Supply Chain Assurance at BMW Group, told German logistics newspaper, *DVZ* (Bottler 2016).

Considering the characteristics of the modern supply chain and requirements for flexibility, reliability, speed, security, and robustness, the supply chain industry has become highly dependent on ICT. Without data management capabilities, modern supply chains cannot be managed anymore. According to the 2013 survey by BVL International, 60% of survey respondents indicated they are planning to invest in big data analytics within the next 5 years (BVL International 2013).

Supply chain innovation, design, and planning can benefit from methodologies originating in the world of software engineering, such as SCRUM, that provide a simple set of rules for enabling more effective team collaboration on highly complex projects. The SCRUM framework places emphasis on specific intermediate results, produced in short, usually weekly, intervals. Instead of abstract analysis and process definitions, SCRUM is designed to help define requirements and find solutions through gradual optimization and approximation. The resulting solutions are often highly flexible, resilient, and lean.

In the digital world, the entire supply and value chain needs to be reengineered; old practices and business models will disappear, and new ones will rise. Contracts will be concluded on collaboration platforms; products will be manufactured using 3D printing technology; the supply and value chain will be increasingly planned, managed, and refined with the help of logistics control towers – the synonym for supply chain visibility across the system. New partnerships are emerging, such as between supply chain solution provider *Panalpina* and 3D printing specialist *Shapeways* or between the software company *SAP* and the delivery leader *UPS* (Air Cargo News 2016). Such partnerships signal the beginning of an industrial realignment, one that will soon see 3D printing incorporated into a seamless supply and manufacturing process from order to delivery.

In April 2015, the European Commission decided to establish the Digital Transport and Logistics Forum (DTLF) to support the digitalization of the freight transport and logistics industries. The DTLF has focused on issues related to standardization and creating a "climate of trust" through data protection and cybersecurity measures and recognition of e-transport documents (European Commission 2015). The DTLF, whose mandate expires after 3 years, has positioned itself as platform for collaboration and coordination among member states, trade unions, and subject matter experts.

The digital revolution that is transforming the world in general has also created the digital supply chain. The term digital supply chain stands for the technologies, many mentioned and explained above, which help to better design, plan, and manage the flow of goods or orchestrate the participants in the supply chain ecosystem. Digitization changes all processes along the supply and value chain, from the way we design, to additive manufacturing, to e-customs clearance, to e-commerce, to e-logistics. The technologies drive a new era of transformation of business models, and the supply chain sits at its core, overlapping with manufacturing and retail.

1.3.3 Logistics Performance

The performance of the global supply chain relies to a great extent on seamless flow of goods across geographies and borders. Although national governments have to balance security concerns and economic interest, open borders are less of a challenge than one might think. From a governmental perspective, the performance of cross-border supply and value chains is a strategic asset for attracting companies and investors to bring knowledge and capabilities, prosperity, and wealth. As a result, governments deploy supporting infrastructure as well as economic and free trade zones; effective legal frameworks, fiscal policies, and multi- and bilateral trade agreements supplement the effort. The access to domestic markets, i.e., the ability to sell to local customers and consumers, is a critical factor for companies considering investments in new markets, in particular, in times when robots increasingly eliminate labor cost advantages.

Strategic investment in future industries and the gradual phasing out of obsolete sectors should be the ultimate goal of governments: not protecting weak industries but driving the local economy towards its comparative advantages and highest level of competitiveness.

Logistics performance determines to a large extent competitiveness and advantage. The World Bank regularly analyses the operating conditions of the logistics sector in different countries and publishes the Logistics Performance Indicator (LPI) every 2 years. The LPI is based on a worldwide survey of 1000 logistics professionals on the ground (global freight forwarders and express carriers), providing feedback on the logistics *friendliness* of the countries in which they operate. Since transport system efficiency and business performance are closely related, the LPI is a valuable tool for governments to understand where they stand and how they can boost efficiency and performance.

The LPI takes into consideration six core dimensions, including the quality of infrastructure for trade and transport, efficiency of customs processes, quality of logistics services on the ground and by sea, performance of customs brokers, and cost effectiveness of logistics solutions. The index ranges from 1 (worst) to 5 (best) and ranks a total of 160 countries.

In the top 30 of the LPI, we find 22 OECD countries and 14 members of the European Union. China moved from 28th in 2014 to 27th in 2016. India did not make its way into the top 30 in 2016 but has jumped 19 places, improving from 65th in 2014 to 38th in 2016 (The New Indian Express 2016). Between 2007 and 2014, the gap between top and bottom LPI performers had begun narrowing, driven by continuous improvements of the infrastructure and service quality of the logistics sector as well as the customs clearance processes. In 2016, however, there was a convergence in LPI performance near the top, with a widening gap between the high and low performer countries. The Philippines offers an example of how fast the picture can change: the archipelago, which ranked 44th in 2010, fell 13 places to 57th in 2014 and to 71st in 2016 (The World Bank 2016). This drop largely resulted from two factors: (1) growing weakness in transport-related infrastructure and (2) declining quality of logistics services, including transport operators and customs brokers.

Logistically constrained countries, such as landlocked nations without direct access to the oceans and global waterways, regularly struggle with trade and transport facilitation and reforms. Beyond internal political will, the logistically disadvantaged countries, which are often today's suppliers and tomorrow's potential customers, require and deserve attention and support from the international community.

The German Association of Logistics (Bundesvereinigung Logistik – BVL) observed vast variations when comparing logistics cost between different regions and countries: while logistics costs in China accounted for more than 14% of business revenues, in the USA, they account for <6%. There are also substantial differences between industries. The BVL calculated that the share of logistics costs among total costs in the materials/mining industry accounted for 13%; for the machine/plant engineering industry, the figure was just 3% in 2013 (BVL International 2013).

The efficiency of customs processes plays an important part in determining the seamlessness of supply chains. Most manufacturers depend on importing intermediates or semifinished goods. Slow and bureaucratic customs processes increase costs. Thus, efficient and speedy customs clearance is essential for the competitiveness of businesses and nations. Customs authorities are mindful about this need and collaborate with international organizations such as the International Civil Aviation Organization (ICAO), International Maritime Organization (IMO), and Financial Action Task Force (FATF) to balance risks with efficiencies. The ultimate aim is the harmonization and simplification of customs procedures across the globe in the best interest of all stakeholders.

The *FATF* is an intergovernmental body with 36 members, 34 member jurisdictions and two regional organizations, as well as various associate members, among them the USA, Canada, most European countries, Australia, and the Russian Federation. The body was established in 1989 to "set standards and promote effective implementation of legal, regulatory and operational measures for combating money laundering, terrorist financing and other related threats to the integrity of the international financial system" (FATF 2016). The FATF has developed a series of recommendations that are recognized as international standard.

The *IMO* currently has 171 member states and associate members. A predecessor, the Inter-Governmental Maritime Consultative Organization, was founded in 1948 at an international conference in Geneva; 10 years later the IMO started its work. Most of their amendments and conventions deal with safety and environmental issues for commercial seafarers. Important for the logistics sector are the facilitation of international maritime traffic and load lines and the carriage of dangerous goods. The system of measuring the tonnage of ships was revised. The conventions are aimed at simplifying the procedures and formalities to be fulfilled when ships arrive in or depart from ports.

The *International Civil Aviation Organization (ICAO)* was established in 1944 to support the safe, efficient, secure, economically sustainable and environmentally responsible civil aviation sector. The international organization has currently 191 member states and industry groups. In 1995 the General Agreement on Trade in Services (GATS) was passed. At present, the GATS Air Transport Annex covers three so-called "soft" rights, namely, aircraft repair and maintenance, selling and marketing of air transport, and computer reservation system (CRS) services.

Another major stakeholder in the international customs community is the *World Customs Organization (WCO)*, whose membership covers 98% of world trade. Founded in 1952 in Brussels, Belgium, the international organization currently has 180 members, three-quarters of which are developing countries. Their mission is to promote security and facilitate international trade, mainly through simplification and harmonization of customs procedures and assurance of compliance with laws and regulations.

In the USA, the Customs-Trade Partnership Against Terrorism (C-TPAT) aims at securing and expediting supply chain processes. This voluntary program is led by the US Customs and Border Protection (CBP), which was launched in November 2001. The focus of the initiative is on improving the security of private companies' supply chains and mitigating any potential risks related to terrorism. In 2016, the program has 11,325 member companies, 4246 of them being U.S importers. The companies account for over 54% of the value of all merchandise imported into the USA (Supply Chain Security International Inc. 2016). Among the participants are US/Canadian and US/Mexican cross-border highway carriers, Mexican long-haul carriers, Mexican and Canadian manufacturers, third-party logistics (3PL) providers, and exporters. Companies wishing to achieve C-TPAT certification must have a documented process for mitigating risk throughout their international supply chain. This allows companies to be considered low risk, resulting in expedited processing of their cargo, including fewer customs inspections. Various programs around the world have mutual recognition with C-TPAT including the New Zealand Customs Service's Secure Export Scheme (SES) program, the European Union and Japan's Authorized Economic Operator (AEO) program, and – most recently in 2014 – Israel's, Mexico's, and Singapore's customs programs.

1.3.3.1 Trade Facilitation Efforts

In 2016, U.S President Barack Obama signed the Trade Facilitation and Trade Enforcement Act of 2015, which aims at streamlining U.S customs procedures even further and reducing shipping costs for transport companies. One of the main elements is the provision to quadruple the de minimis threshold on inbound shipments from $200 to $800. The de minimis is defined as valuation ceiling for imports, below which no duty or tax is charged and clearance procedures are minimal. This bill also supports the creation of a single-window system, making the communication and exchange of information easier between all parties involved.

In 2015, the USA's largest trading partner in total trade was China, followed by Canada, Mexico, Japan, and Germany (United States Census Bureau 2015). Canada and Mexico are the USA's two largest export markets. Since 1994, the three countries have implemented the *North American Free Trade Agreement (NAFTA)*. It is worth noting that nearly 60% of U.S imports from Mexico and Canada are used in the production of made-in-America goods and services (Office of the United States Trade Representative 2015).

NAFTA, which encompasses a total population of 450 million, is the world's largest free trade agreement, as measured by the combined GDP ($20.08 trillion) of the USA ($16.72 trillion), Canada ($1.518 trillion), and Mexico ($1.845 trillion). In terms of economic output, NAFTA generates more than the 28 countries in the European Union. Between 1993 and 2014, trade between the three NAFTA members quadrupled, from $297 billion to $1.2 trillion. U.S exports grew from $142 billion to $552.3 billion. That means 34% of all U.S exports went to its trading partners, Mexico and Canada (Amadeo 2016). The agriculture, automotive, service, textile, and apparel industries have all been major beneficiaries. For example, U.S food exports to Canada and Mexico grew 156% since the signing of the agreement, and exports to the rest of the world only 65%. For the American government, NAFTA is equally important to the hydrocarbon trade. The USA imports hydrocarbons mainly from Mexico and Canada, which reduces U.S reliance on oil imports from the Middle East and Venezuela.

The USA and Europe are currently negotiating the *Transatlantic Trade and Investment Partnership (TTIP)*. The European Commission estimates 90% of global demand will be generated outside Europe over the next 10 to 15 years (European Commission 2016). The American government considers the TTIP a companion agreement to the *Trans-Pacific Partnership (TPP)*, agreed upon in October 2015, which is meant to eliminate or reduce tariff and nontariff barriers between the USA, Brunei, Chile, New Zealand, Singapore, Australia, Canada, Japan, Malaysia, Mexico, Peru, and Vietnam. The TPP is meant to promote the development of production and supply chains and facilitate seamless trade. Furthermore, TPP is intended as a platform for enabling greater regional economic integration, with plans to include additional countries across the Asia-Pacific region (Office of the United States Trade Representative 2015).

All in all, the USA has free trade agreements with 20 countries, among them the *Dominican Republic-Central America FTA (CAFTA-DR)*, which was signed between the USA, Costa Rica, El Salvador, Guatemala, Honduras, Nicaragua, and the Dominican Republic. Central America and the Dominican Republic represent the third largest US export market in Latin America, behind Mexico and Brazil. Total two-way goods trade between the USA and the six CAFTA-DR partners has increased over 71% since the agreement took force, from $35 billion in 2005 to $60 billion in 2013 (Office of the United States Trade Representative 2016a, b).

Another important agreement is the *Trade and Investment Framework Agreement (TIFA)*, which provides a strategic framework for governments to discuss and resolve trade and investment issues at an early stage. Currently, the USA has TIFAs with nearly 70 countries at different levels of development trade and investment.

The EU, whose largest trading partner is the USA, currently holds 15 free trade agreements with 31 countries (BMWI 2014). In early 2016, Canada and the EU agreed a new trade partnership, called the *Comprehensive Economic and Trade Agreement (CETA)*. According to estimates by a joint Canada-EU study, CETA is likely to boost both economies by 12 billion euros (BMWI 2016) and CAD $12 billion accordingly per year (Government of Canada 2015).

Another free trade zone is expected to be established in the East with the signing of the *Regional Comprehensive Economic Partnership (RCEP)*, linking the ten member states of the Association of Southeast Asian Nations (ASEAN) with Australia, China, India, Japan, South Korea, and New Zealand. RCEP negotiations were formally launched in November 2012 and are expected to be finalized by the end of 2016. Including more than 3 billion people, the RCEP reaches a combined GDP of about $17 trillion, a trade volume of $10.6 billion, and will account for about 30% of global trade (Tang 2015).

1.4 Undesired Consequences

Production does not come without controversy. For example, the production of one cotton shirt requires about 2700 l of water (World Wildlife Fund 2013). Environmentalist argues that the shrinking of the Aral Sea to just 15% of its original size in the past 20 years is mostly caused by cotton farming. Where there was once water, there is a now salt desert, contaminated with pesticide and toxic dust. What little remains of the Aral Sea is now almost completely lifeless. Only about 27% of cotton grown in Uzbekistan is rain-fed. The rest is produced in irrigated fields, which leads to major water loss due to seepage, evaporation, and poor water management. In Uzbekistan, it is estimated that 60% of water is lost through its irrigation system. India is said to use huge amounts of pesticides on conventionally grown (i.e., nonorganic) cotton. Cotton in India accounts for just 5% of agricultural land, yet for 54% of annual pesticide use. In the wake of this kind of contamination, DNA mutation and major health problems are expected to afflict ever greater segments of the local population (people & planet 2015).

Gap has published a list of nearly 50 countries where their primary suppliers are located. *H&M* shared the names and locations of about 800 factories in Asia and Europe, covering 95% of its supplier base. With suppliers typically spread around the world, cotton is shipped from one continent to the other to processing or manufacturing plants, turned into clothing, and shipped then to consumers all across the globe – with high energy consumption and a large carbon footprint. All in all, a shirt can travel 16,500 miles from the cotton fields to the shop (Green 2015).

The deadly Rana Plaza fire and building collapse in Bangladesh has increased awareness about the dangerous conditions within the garment industry. Rana Plaza was by no means the first large-scale fire and collapse at a garment factory in Bangladesh. The Bangladesh Accord Foundation reports that garment factory fires and collapses in Bangladesh have killed at least 1800 workers since 2005 (Bangladesh Accord Foundation 2015). More than \$280 million have been committed by the international community to raise standards in the industry in Bangladesh (White 2015). However, a study conducted by New York University found that, of the 3425 site inspections concluded in Bangladesh, only eight factories have addressed violations to pass final inspection.

The environmental and social dimensions of international trade are growing increasingly important and exerting substantial pressures on specific sectors of the global economy. Greenpeace, the nongovernmental environmental advocacy group, has been claiming for years that the fashion industry is using substances harmful to the environment and human health. According to Greenpeace, 64% of urban underground water in China and 80% of water in Jakarta's main river is heavily polluted (Greenpeace 2015). Fast fashion retailers, such as *Primark* and *H&M*, as well as labels such as Levi's and Burberry, claim to have banned toxic chemicals from production. In March 2015, Greenpeace announced that 10% of the global retail fashion industry has committed to eliminating toxic chemicals. In July 2015, thanks in part to pressure from Greenpeace, EU member states voted to ban the toxic chemical NPE from textile imports.

Environmental consequences connected with exploration and extraction of rare earth elements are also significant. The 17 elements are used in the production of electronics, catalytic converters, infrared lasers, and medical devices, among other things. Mining, refining, and disposal of rare earth elements cause major environmental burdens; tailings are a by-product of mining rare earth elements of particular concern, because they can be absorbed into the surrounding groundwater. When processed, rare earth elements are often laced with radioactive materials such as thorium and carcinogenic sulfate, in addition to ammonia and hydrochloric acid. Processing one ton of rare earth elements produces 2000 t of toxic waste.

Until 1948, most of the world's rare earth elements were sourced from India and Brazil. After large veins of monazite were discovered in the 1950s, South Africa became the world's leading producer of rare earth elements. From the 1960s to the 1980s, California took the top spot with rare earth element production from the Mountain Pass mine. Today, while Indian and South African deposits still yield some rare earth concentrates, they are dwarfed by the scale of Chinese production: in 2014, China was responsible for 85% of the world's rare earth element production. Most of the Chinese production is concentrated near Baotou, a city in China's Inner Mongolia Autonomous Region. Baotou's rare earth enterprises produce ten million tons of wastewater per year.

New demand for rare earth elements has strained supply, and there is growing concern that the world may soon face a shortage. Unless major new sources of rare earth elements are unlocked, global demand is expected to exceed total supply by 40,000 t annually within a few years (Kaiman 2014). In the meantime, the world's first hybrid car engine that doesn't use heavy rare earth metals has already been produced (The Wall Street Journal 2016).

The growing pressure on the earth is becoming more and more pronounced; the earth's population is forecasted to rise from 7 billion in 2016 to an estimated nine billion in 2050. Consequently, governments, businesses, and citizens are required to work towards innovative solutions to ensure that resource demand is met without jeopardizing the supply and environment of future generations.

Literature

3Dprinting.com. (2015). *What is 3D printing?* Accessed March 15, 2016, from http://3dprinting.com/what-is-3d-printing/

Air Cargo News. (2016, July 29). *3D printing, a supply chain challenge.* Accessed September 15, 2016, from http://www.aircargonews.net/news/single-view/news/3d-printing-a-supply-chain-challenge.html

Amadeo, K. (2016, March 23). *Advantages of NAFTA.* about.com. Accessed January 04, 2016, from http://useconomy.about.com/od/tradepolicy/p/NAFTA_Advantage.htm

Associated Press. (2015, July 24). *Fiat Chrysler recalls 1.4m vehicles in wake of Jeep hacking revelation.* The Guardian. Accessed September 15, 2016, from https://www.theguardian.com/business/2015/jul/24/fiat-chrysler-recall-jeep-hacking

Association for Operations Management (APICS). (2015). *Exploring the big data revolution* (p. 8)

Bangladesh Accord. (2015). Accessed February 24, 2016, from http://bangladeshaccord.org/bangladesh/

Banker, S. (2016, April 7). *A fresh look at supply chain visibility.* Forbes. Accessed April 15, 2016, from http://www.forbes.com/sites/stevebanker/2016/04/07/a-fresh-look-at-supply-chain-visibility/#34e69a9842ca

BBC News. (2014, September 15). *How far do US-EU sanctions on Russia go?* Accessed December 14, 2015, from http://www.bbc.com/news/world-europe-28400218

Bloomberg News. (2016, January 28). *China's CRRC said in hyperloop talks as musk vision moves closer.* Bloomberg. Accessed September 15, 2016, from http://www.bloomberg.com/news/articles/2016-01-28/china-s-crrc-said-in-hyperloop-talks-as-musk-vision-moves-closer

Bolle, M. J., & Williams, B. R. (2013, November 12). *U.S. foreign-trade zones: Background and issues for congress.* Congressional Research Service. Accessed November 12, 2015, from https://www.fas.org/sgp/crs/misc/R42686.pdf

Bottler, S. (2016, March 18). *BMW macht Versorgung sicherer* (p. 5). Hamburg: Deutsche Verkehrs-Zeitung.

Bundesministerium für Wirtschaft und Energie. Außenwirtschaft. (2014). *Freihandelsabkommen der EU.* Accessed January 14, 2016, from http://www.bmwi.de/DE/Themen/aussenwirtschaft, did=666642.html

Bundesministerium für Wirtschaft und Energie. Außenwirtschaft. (2016). *European trade policy.* Accessed January 13, 2016, from http://www.bmwi.de/EN/Topics/Foreign-trade/Trade-policy/european-trade-policy.html

BVL International. (2013). *Trends and strategies in logistics and supply chain management* (pp. 20 and 50).

Campbell, M., & Webb, A. (2014, July 30). *Siemens to BP prepare for downward russia business spiral.* Bloomberg. Accessed March 30, 2016, from http://www.bloomberg.com/news/articles/2014-07-29/siemens-to-bp-prepare-for-downward-russia-business-spiral

Christopher, M., & Peck, H. (2004). Building the resilient supply chain. *International Journal of Logistics Management, Cranfield School of Management, 15*(2), 1–13. Accessed November

12, 2015, from https://dspace.lib.cranfield.ac.uk/bitstream/1826/2666/1/Building percent20the percent20resilient percent20supply percent20chain-2003.pdf

CIPS/Dun & Bradstreet. (2016). *CIPS Risk Index—quarterly report Q1 2016*

Cuthbertson, A. (2016, June 23). *Vladimir putin imagines hyperloop silk road*. Accessed September 14, 2016, from http://europe.newsweek.com/hyperloop-vladimir-putin-silk-road-473573?rm=eu

Dattel, E. R. (2015). *Cotton in a global economy*. Accessed December 15, 2015, from http://mshistorynow.mdah.state.ms.us/articles/161/cotton-in-a-global-economy-mississippi-1800-1860

Deloitte. (2016, April 1). *Assessing supply chain risk: 30 percent experienced supply chain fraud in past year*. Accessed April 15, 2016, from http://de.slideshare.net/DeloitteUS/assessing-supply-chain-risk-30-percent-experienced-supply-chain-fraud-in-past-year

Deryabina, A. (2015, August 18). *Which countries have benefited from Russia's food embargo?* Russia Beyond The Headlines (RBHT). Accessed March 30, 2016, from http://rbth.com/business/2015/08/18/which_countries_have_benefited_from_russias_food_embargo_48587.html

DF Deutsche Forfait AG. (2014, October 16). *Ad hoc-Mitteilung gem. § 15 WpHG: DF Deutsche Forfait AG wurde von der OFAC Sanktionsliste entfernt*. Accessed April 20, 2016

DHL Global. (2014, February 13). *DHL uses big data for risk mitigation in logistics*. Bonn. Accessed March 21, 2016, from http://www.dhl.com/en/press/releases/releases_2014/logistics/dhl_uses_big_data_for_risk_mitigation_in_logistics.html

Earls, A., & Baya, V. (2014). *Technology forecast: The future of 3D-printing—The road ahead for 3D printers* (Issue 2, p. 2). PricewaterhouseCoopers

European Commission, Mobility and Transport. (2015, October 28). *Commission launches forum on digital transport and logistics*. Accessed April 10, 2016, from http://ec.europa.eu/transport/media/events/2015-07-01-dtlf_en.htm

European Commission—Trade. (2016). *Trade negotiations in a nutshell*. Accessed September 15, 2016, from http://ec.europa.eu/trade/policy/countries-and-regions/agreements/

European Union, Newsroom. (2015). *EU sanctions against Russia over Ukraine crisis*. Accessed December 14, 2015, from http://europa.eu/newsroom/highlights/special-coverage/eu_sanctions/index_en.htm

Evenett, S. J., & Fritz, J. (2015, November 12). *The tide turns? Trade, protectionism, and slowing global growth*. Centre for Economic Policy Research, CEPR Press. Available at http://www.globaltradealert.org/sites/default/files/GTA18%20The%20Tide%20Turns.pdf

EWE. (2015, April 15). *EWE acquires VNG shares from Gazprom*. Oldenburg. Accessed March 31, 2016, from https://www.ewe.com/en/media/press-releases/2015/04/ewe-acquires-vng-shares-from-gazprom-ewe-ag

Farrington, D. (2016, May 11). *Elon Musk's hyperloop dream has its 1st public demo*. NPR. Accessed September 15, 2016, from http://www.npr.org/sections/thetwo-way/2016/05/11/477645103/elon-musks-hyperloop-dream-is-about-to-have-its-1st-public-demo

FATF. (2016). http://www.fatf-gafi.org/about/

Gartner Inc. (2014, November 11). *Gartner says 4.9 billion connected 'things' will be in use in 2015*. Gartner. Accessed September 14, 2016, from http://www.gartner.com/newsroom/id/2905717

Ghosh, N. (2016, January 21). *China's dream of rail link to S-E Asia coming true*. The Straits Times. Accessed April 06, 2016, from http://www.straitstimes.com/asia/east-asia/chinas-dream-of-rail-link-to-s-e-asia-coming-true

Giffi, C. A., Gangula, B., & Illinda, P. (2014, May 19). *3D opportunity for the automotive industry*. Deloitte University Press. Accessed March 16, 2016, from http://dupress.com/articles/additive-manufacturing-3d-opportunity-in-automotive/

Government of Canada, Global Affairs Canada. (2015). *Canada-European Union: Comprehensive economic and trade agreement (CETA)*. Accessed January 13, 2016, from http://international.gc.ca/trade-agreements-accords-commerciaux/agr-acc/ceta-aecg/understanding-comprendre/brief-bref.aspx?lang=eng

Green, M. (2015, March 18). *Where does your T-shirt come from? Follow its epic global journey.* KQED News. Accessed January 14, 2016, from http://ww2.kqed.org/lowdown/2015/03/18/making-your-t-shirt-a-journey-around-the-world/

Greenpeace. (2015, March 19). *What the fashion industry looks like after 4 years of Detox.* Accessed January 03, 2016, from http://www.greenpeace.org/international/en/news/Blogs/makingwaves/detox-catwalk/blog/52356/

Grothaus, M. (2016, January 22). *Two companies are racing to build the first hyperloop.* Fast Company & Inc. Accessed September, 14, 2016, from http://www.fastcompany.com/3055801/fast-feed/two-companies-are-racing-to-build-the-first-hyperloop

Hawkins, A. J. (2016, January 20). *Hyperloop transportation is about to break ground on its first test track.* The Verge. Accessed September 15, 2016, from http://www.theverge.com/2016/1/20/10800654/hyperloop-test-track-construction-plans-quay-valley-elon-musk

Higgins, T. (2014, December 16). *Apple stops online sales in Russia over ruble fluctuations.* Bloomberg. Accessed March 30, 2016, from http://www.bloomberg.com/news/articles/2014-12-16/apple-stops-online-sales-in-russia-over-ruble-fluctuations [Higgins, T. (2015, February 2). *Apple not ditching Russian market—Retailers.* Accessed March 30, 2016, from https://www.rt.com/business/228583-apple-products-russia-continue/]

IDC. (2016). *Smartphone OS market share 2015 Q2.* Accessed January 14, 2016, from http://www.idc.com/prodserv/smartphone-os-market-share.jsp

Institut der deutschen Wirtschaft. (2015). *Globale Kräfteverschiebung. Kräfteverschiebung in der Weltwirtschaft—Wo steht die deutsche Industrie in der Globalisierung?* (p. 13). Köln

Intuit Inc. (2014, August 7). *Small business success in the cloud.* LinkedIn SlideShare. Accessed September 15, 2016, from http://www.slideshare.net/IntuitInc/future-of-smb-for-slidesharev2/1

Jackson, M. O., & Nei, S. M. (2015, June). *Networks of military alliances, wars, and international trade* (Draft). Stanford, CA: Department of Economics, Stanford University

Jamison, P., & Kain, H. (2016, April). *Managing supply chain risk in a flat world* (p. 9). ALOM & MetricStream

Johnston, M. (2015, August 05). *How Russian sanctions impact western companies.* Investopedia. Accessed March 30, 2016, from http://www.investopedia.com/articles/investing/080515/how-russian-sanctions-impact-western-companies.asp

Kaiman, J. (2014, March 20). *Rare earth mining in China: The bleak social and environmental costs.* The Guardian. Accessed January 03, 2016, from http://www.theguardian.com/sustainable-business/rare-earth-mining-china-social-environmental-costs

Kane, G. C., Palmer, D., Nguyen Phillips, A., Kiron, D., & Buckley, N. (2016, July 26). *Aligning the organization for its digital future.* MIT Sloan Management Review & Deloitte University Press. Accessed September 15, 2016, from http://sloanreview.mit.edu/projects/aligning-for-digital-future/

Keppner, K. (2015, November 30). *Micro-risks—A new perspective on supply chain risk management.* Accessed April 11, 2016, from http://www.inventory-and-supplychain-blog.com/micro-risks-a-new-perspective-on-supply-chain-risk-management/

Kuchma, A. (2015, January 29). *Russia is facing record capital investment outflow.* Russia Direct. Accessed March 23, 2016, from http://www.russia-direct.org/russian-media/russia-facing-record-capital-and-investment-outflow

Lehmacher, W. (2016, July 20). *How smart packaging can save lives.* World Economic Forum. Accessed September 15, 2016, from https://www.weforum.org/agenda/2015/07/how-smart-packaging-can-save-lives

Lehmacher, W., & Padilla-Taylor, V. (2015, November). The new silk road—Idea and concept. In *IPSW strategy series: Focus on defense and international security* (Issue No. 390)

Lipton, J. (2016, March 7). *Hyperloop will be here in 2020 and the impact will be huge.* CNBC. Accessed September 14, 2016, from http://www.cnbc.com/2016/03/07/hyperloop-will-be-here-in-2020-and-the-impact-will-be-huge.html

Macguire, E., & Anderson, B. (2013, June 27). *'Silk Road' railways link Europe and Asia*. CNN. Accessed April 06, 2016, from http://edition.cnn.com/2013/06/27/business/silk-railroad-trading-network/

Manenti, P. (2016, August 2). *The burning platform for future competitiveness*. SCM World. Accessed September 15, 2016, from http://www.scmworld.com/burning-platform-future-competitiveness/

Mangasarian, L. (2015, July 29). *Claas tractor bets on Russia expansions as others flee country*. Bloomberg. Accessed March 30, 2016, from http://www.bloomberg.com/news/articles/2015-07-28/claas-tractors-bets-on-russia-expansion-as-others-flee-country

Manyika, J., Lund, S., Bughin, J., Woetzel, J., Stamenov, K., & Dhingra, D. (2016, February). *Digital globalization: The new era of global flows*. McKinsey Global Institute. Accessed March 15, 2016, from http://www.mckinsey.com/business-functions/mckinsey-digital/our-insights/digital-globalization-the-new-era-of-global-flows

National Museums Liverpool, Merseyside Maritime Museum. (2015). *Exhibitions, 100 percent cotton, cotton traders, trading rules*. Accessed December 15, 2015, from http://www.liverpoolmuseums.org.uk/maritime/exhibitions/cotton/traders/trading-rules.aspx

Nicas, J. (2016, January 21). Hyperloop race picks up speed. *The Wall Street Journal*. Accessed September 14, 2016, from http://www.wsj.com/articles/hyperloop-race-picks-up-speed-1453426042

Nurshayeva, R. (2013, June 10). *Kazakhs launch 'Silk Road' China-Europe rail route*. Reuters. http://www.reuters.com/article/us-kazakhstan-railway-idUSBRE9590GH20130610

Office of the United States Trade Representative. (2015). *Free trade agreement, North American Free Trade Agreement (NAFTA)*. Accessed December 11, 2015, from https://ustr.gov/trade-agreements/free-trade-agreements/north-american-free-trade-agreement-nafta

Office of the United States Trade Representative. (2016a). *Summary of the trans-Pacific partnership agreement*. Accessed January 13, 2016, from https://ustr.gov/about-us/policy-offices/press-office/press-releases/2015/october/summary-trans-pacific-partnership

Office of the United States Trade Representative. (2016b). *Trade agreements/free trade agreements, Dominican Republic-Central America FTA*. Accessed January 13, 2016, from https://ustr.gov/trade-agreements/free-trade-agreements/cafta-dr-dominican-republic-central-america-fta

O'Toole, J. (2014, July 29). *Amazon launches 3D-printing store*. CNN Money. Accessed March 16, 2016, from http://money.cnn.com/2014/07/29/technology/innovationnation/amazon-3d-printing/

Pakistan Times. (2016). *China-Pakistan Economic Corridor (CPEC)*. Accessed April 07, 2016, from http://www.pakistantimes.com/topics/china-pakistan-economic-corridor/

people & planet. (2015). *The cost of cotton: Dirty cotton*. Accessed December 15, 2015, from https://peopleandplanet.org/redressfashion/briefing/dirty

Peters, A. (2015, July 29). *What if the milk container told you when the milk was bad?* Fast Company & Inc. Accessed Sepetember 15, 2016, from https://www.fastcoexist.com/3049092/what-if-the-milk-container-told-you-when-the-milk-was-bad

Port of Rotterdam Authority. (2016, February 11). *RDM Rotterdam to acquire a 'Fieldlab' with 3D metal printers*. Accessed March 16, 2016, from https://www.portofrotterdam.com/en/news-and-press-releases/rdm-rotterdam-to-acquire-a-fieldlab-with-3d-metal-printers

Porter, M. E. (1985). *Competitive advantage: Creating and sustaining superior performance*. New York, NY: Simon and Schuster. Retrieved September 9, 2013

Railway Technical Web Pages. (2016). *Railway finance*. Accessed September 14, 2016, from http://www.railway-technical.com/finance.shtml

Rapoza, K. (2015, March 17). Only slight drop in U.S.-Russia trade since sanctions. *Forbes Magazine*. Accessed December 14, 2015, from http://www.forbes.com/sites/kenrapoza/2015/03/17/only-slight-drop-in-u-s-russia-trade-since-sanctions/

Raven, L. (2012, June 20). *Shapeways hits one million 3D printed creation.* TG daily. Accessed March 16, 2016, from http://www.tgdaily.com/general-science-brief/64170-shapeways-hits-one-million-3d-printed-creations

Rhenus. (2015, May 6). *Logistics specialists register increases in flows of goods from Asia heading for Russia.* Holzwickede, Germany. Accessed March 31, 2016, from http://www.rhenus.com/en/infocenter/press/single-news/article/logistiker-verzeichnen-verstaerkte-warenstroeme-aus-asien-in-richtung-russland/

Sareen, H. (2015, April 7). *The 6 biggest takeaways from Amazon's dash button.* ClickZ. Accessed September 15, 2016, from https://www.clickz.com/the-6-biggest-takeaways-from-amazons-dash-button/26534/

Simchi-Levi, D. (2015, June 9). Find the weak link in your supply chain. *Harvard Business Review.* Accessed April 07, 2016, from https://hbr.org/2015/06/find-the-weak-link-in-your-supply-chain

Srinivasan, V., & Bassan, J. (2012, December 7). *Manufacturing the future: 10 trends to come in 3D printing.* Forbes. Accessed March 16, 2016, from http://www.forbes.com/sites/ciocentral/2012/12/07/manufacturing-the-future-10-trends-to-come-in-3d-printing/#5a90905e2ba3

Statista, The Statistics Portal. (2015). *Cotton production by top country.* Accessed December 16, 2015, from http://www.statista.com/statistics/263055/cotton-production-worldwide-by-top-countries/

Statista, The Statistics Portal. (2016). *Production of cells phones in China.* Accessed January 14, 2016, from http://www.statista.com/statistics/226434/production-of-cell-phones-in-china-by-month/

Supply Chain @MIT. (2013, October 3). *Bringing the cloud down to earth.* Accessed March 17, 2016, from http://supplychainmit.com/2013/10/03/bringing-the-cloud-down-to-earth/

Supply Chain Security International Inc. (2016). *What is C-TPAT?* Accessed September 15, 2016, from http://c-tpat.com/what-is-ctpat/

Tang, S. K. (2015, October 14). *RCEP: The next trade deal you need to know about.* CNBC.

The Economist. (2014, July 12). *The flow of things.* Accessed September 14, 2016, from http://www.economist.com/news/china/21606899-export-superpower-china-suffers-surprisingly-inefficient-logistics-flow-things

The Ethical Consumer. (2016). *History of successful boycotts.* Accessed March 15, 2016, from http://www.ethicalconsumer.org/boycotts/successfulboycotts.aspx

The New Indian Express. (2016, July 2). *India jumps 19 places in WB logistics index.* Accessed Sepetmber 15, 2016, from http://www.newindianexpress.com/business/news/India-jumps-19-places-in-WB-logistics-index/2016/07/02/article3509529.ece

The Siberian Times. (2014a). Accessed April 06, 2016, from http://siberiantimes.com/business/casestudy/news/new-chinese-plan-sees-a-rail-link-via-siberia-from-beijing-railway-to-the-us-and-canadaat-bullet-speed/

The Siberian Times. (2014b, May 18). *New Chinese plan sees a rail link via Siberia from Beijing to the US and Canada...at bullet speed.*

The World Bank. (2016). *International LPI global ranking.* Accessed September 15, 2016, from http://lpi.worldbank.org/international/global/2010

Thumm, J. (2016, February 11). *3D printing to change how we shop for shoes.* PowerRetail. Accessed March 16, 2014, from http://www.powerretail.com.au/multichannel/3d-printing-shoes/

Trade Act of 1974 [Public Law 93–618, as amended] [As Amended Through P.L. 114–27, Enacted June 29, 2015]. Accessed November 12, 2015, from https://legcounsel.house.gov/Comps/93-618.pdf

Trading Economics. (2016). *Iran GDP, 1965-2016.* Accessed April 20, 2016, from http://www.tradingeconomics.com/iran/gdp

T-Systems. (2016). *Big data in retail: Pricing—A key retail process.* Accessed March 21, 2016, from http://www.t-systemsus.com/aboutsystems/big-data-in-retail-pricing-a-key-retail-process/1023240

U.S. Chamber of Commerce. (2015). *Global supply chain, customs and trade facilitation.* Accessed November 12, 2015, from https://www.uschamber.com/issue-brief/global-supply-chain-customs-and-trade-facilitation

U.S. Patent & Trademark Office. (2015). Patent Number: 8412588. Accessed March 16, 2016, from http://appft1.uspto.gov/netacgi/nph-Parser?Sect1=PTO2&Sect2=HITOFF&p=1&u=%2Fnetahtml%2FPTO%2Fsearch-bool.html&r=1&f=G&l=50&co1=AND&d=PG01&s1=20150052024.PGNR.&OS=DN/20150052024RS=DN/20150052024

Underwriters Laboratories Inc. (2011). *The life cycle of materials in mobile phones.* Available at http://services.ul.com/wp-content/uploads/sites/4/2014/05/ULE_CellPhone_White_Paper_V2.pdf

United States Census Bureau. (2015). *Foreign trade.* Accessed January 13, 2016, from https://www.census.gov/foreign-trade/statistics/highlights/top/top1511yr.html

US Department of Treasury. (2015). *Sanctions programs and country information.* Accessed December 15, 2015, from https://www.treasury.gov/resource-center/sanctions/Programs/Pages/Programs.aspx

VerkehrsRundschau. (2015, January 21). *Russlandkrise trifft viele Spediteure.* Accessed March 30, 2016, from http://www.verkehrsrundschau.de/russlandkrise-trifft-viele-spediteure-1590955.html

VerkehrsRundschau. (2016). *Polen und Russland einigen sich* (p. 8). #8/2016

Vorotnikov, V. (2015, June 8). *Russia restricts South American meat imports.* Global Meat News. Accessed March 30, 2016, from http://www.globalmeatnews.com/Industry-Markets/Russia-restricts-South-American-meat-imports

Wall Street Journal. (2016, July 18). *China's rare-Earths bust.* Accessed September 15, 2016, from http://www.wsj.com/articles/chinas-rare-earths-bust-1468860856

Watson, B. (2015, January 6). *Do boycotts really work?* The Guardian. Accessed March 15, 2016, from http://www.theguardian.com/vital-signs/2015/jan/06/boycotts-shopping-protests-activists-consumers

White, G. B. (2015, December 17). *Are factories in Bangladesh any safer now?* The Atlantic. Accessed February 15, 2016, from http://www.theatlantic.com/business/archive/2015/12/bangladesh-factory-workers/421005/

Wohlers Report. (2015). ISBN:978-0-9913332-1-9. Available at http://www.wohlersassociates.com

Woods, R. (2016, April 6). *The Iranian Conundrum: How sanctions removal affects global logistics.* AirCargo World. Accessed April 20, 2016, from http://aircargoworld.com/the-iranian-conundrum-how-sanctions-removal-affects-global-logistics/

World Wildlife Fund. (2013). *The impact of a cotton T-shirt.* Accessed December 15, 2015, from http://www.worldwildlife.org/stories/the-impact-of-a-cotton-t-shirt

Xerox® Supplies for All-Inclusive Contracts Brochure. (2014). *How we simplify your supplies management.* Xerox Corporation. Available at https://www.xerox.com/downloads/europe/a/ASR_Customer_Benefits_UK.pdf

Global Supply Chain: Golden Gate or Sword of Damocles?

2

Early on, Western economies largely benefitted from international trade, which was made possible by global supply and value chains. In the last decades, the developing world has significantly increased its participation in global trade. The balancing of the developments to avoid economic losers and geopolitical frictions along the way is one of the major concerns and challenges of modern times.

In 2013, *The Economist* wrote that nearly one billion people have been taken out of extreme poverty over 20 years (The Economist 2013). On October 4, 2015, the World Bank announced that global poverty would fall below 10% for the first time on record (The World Bank 2015). This development reflects the entry of China, India, and other developing countries into global procurement, manufacturing, and distribution systems. Still, many developing countries remain on the margins of global markets and require assistance to strengthen supply chains while elevating domestic skill levels and capacity.

The fluid flow of goods and services across national borders requires the support of domestic populations in the West and the East. The recent UK vote to leave the EU shows a risk in quite the opposite direction – towards disintegration. Governments must ensure that their economic policies ensure inclusive growth. Joseph E. Stiglitz, Professor, Columbia University, writes: "Opinion polls, including a careful study by Stanley Greenberg and his associates for the Roosevelt Institute, show that trade is among the major sources of discontent for a large share of Americans. Similar views are apparent in Europe" (Democracy Corps 2016).

Although globalization has created many more economic winners than losers, early proponents seemed to promise that all would benefit from globalization. Branko Milanovic's analysis shows that working classes in advanced countries, as represented by those who supported Brexit in the UK, have gained little or nothing. Joseph Stiglitz continues: "if globalization is to benefit most members of society, strong social-protection measures must be in place. The Scandinavians figured this out long ago; it was part of the social contract that maintained an open society, open to globalization and changes in technology" (Lidegaard 2016).

© Springer International Publishing AG 2017

W. Lehmacher, *The Global Supply Chain*, Management for Professionals, DOI 10.1007/978-3-319-51115-3_2

Global disintegration and isolationist tendencies, such as Brexit and the proliferation of border fences in Eastern Europe, point to a growing sense of insecurity and economic uncertainty. These factors play an important role in the discussions around major trade agreements, such as TTIP and TPP. Of course, these issues directly affect the performance of global supply chains.

The invisible conveyer belt of the global economy creates new industries, companies, and jobs, in everything from farming and mining, manufacturing, and processing to distribution, reuse, remanufacturing, and recycling. Along the process, capital is raised, investments made, revenues generated, salaries paid, profits made, and investors repaid. At the same time, the global supply chain allows for moving production away from less favorable locations. Consequences can include job losses and social tensions.

Economic growth is not a given. Lasting economic success, whether in the East or the West, requires continuous transition through policy and regulatory reform and investment. Regardless of a country's current stage of economic development, governments need to invest in elevating their workforce and infrastructure to ensure participation in the global value chain. Governments must assist their constituents whose jobs have been displaced, providing training and employment alternatives either in the same industry, in a new industry, or with the state itself. What role a country can play and where a national economy can participate in global value chains depends on many local conditions, such as political stability, maturity of digital and transport infrastructure, availability of a skilled workforce, efficiency of customs clearance, taxation, and regulation.

2.1   The Rise of the Emerging Markets

Over the past decades, many emerging markets achieved significant and prolonged periods of growth, becoming new sources of economic opportunity for the global economy. Although the pace of growth in emerging markets has recently slowed, the journey has not ended. According to Roland Berger, by 2030, Asia's economy will be larger than that of the USA and the EU combined, with the region's share of global GDP climbing from 30% to more than 40% in just 15 years. This will also fuel growth in the logistics industry. The contract logistics market in China, India, Indonesia, and the Philippines is expected to record an annual growth rate >10% (Roland Berger 2014).

While some emerging economies have been able to resume the path to globalization and more inclusive growth, others have faced more persistent challenges. The analyses of emerging economies provide valuable insights on the importance and dynamics of global connectedness and local reforms and transformation.

2.1.1   The BRICS

The Big Four developing economies, Brazil, China, the Russian Federation, and India, account for $14.8 trillion of combined GDP (Statista 2016), while the US total GDP reaches $17.9 trillion. Recently, the bloc has taken a tumble: the economies of Brazil and Russia are substantially underperforming relative to China and India.

2.1.1.1 Brazil

Brazil's economy is the largest in South America. The GDP of Brazil represents 3.78% of the global economy. From 2000 to 2011, Brazil was one of the fastest-growing large economies in the world. Since then, Brazil's economic activity has stalled, with the currency declining 60% in USD terms since 2011 and with double digit inflation (Biller 2016). According to the World Bank, the poverty headcount ratio at the national poverty line as a percentage of population declined steadily from 13.3% in 2019 to 7.4% in 2014 (World Bank 2016a, b, c, d, e, f, g).

Brazil's growth trajectory correlates closely with global demand for commodities. In 2014, raw materials accounted for half of the value of Brazilian exports, most of which were destined for the USA and China. During 2015, Brazil experienced major declines in export revenue from aluminum (-31.1%), iron ore (-36%), gasoline (-36.5%), and oil (-39.4%) (Trading Economics 2016a). The reason for the negative economic development is an erratic economic policy, serious infrastructure deficiencies, high taxes, and a slow working bureaucracy that hinders investments (Meyer 2014). *Bloomberg* states that the investments that would have made the economy more efficient remained well below half that of China as a percentage of GDP.

Still, Brazil remains an attractive location and trading partner, because of its large domestic market: it's the fifth largest country in the world, by landmass and population. Its offshore oil reserves include the Western Hemisphere's biggest discovery since 1976. Brazil has the second largest iron ore reserves and is the second largest producer of soybeans and the third largest of corn (Bloomberg 2016).

For the USA, Brazil is the 12th largest trading partner, with 1.6% of overall trade. Even though US-Brazilian relations have been strained over the past year as a result of alleged National Security Agency (NSA) activities inside Brazil, the countries continue to engage on topics such as trade, energy, security, racial equality, and the environment. For example, as part of the farm bill reauthorization, Congress approved modifications to the U.S cotton program that could help resolve a long-running trade dispute with Brazil. Congress also continued to support conservation of the Amazon Rainforest, appropriating $10.5 million for environmental programs in the Brazilian Amazon in the Consolidated Appropriations Act, 2014 (Meyer 2014).

Car manufacturers, in particular, are providing jobs for the middle class in Brazil. Manufacturers of cell phones, tablets, and computers are also prominent employers. Leading foreign players have invested in the country: among the ten biggest companies in Brazil are General Motors from the USA and Europe-based

Telefónica, Volkswagen, and Shell. São Paulo has become the international gateway and a popular industrial location. The state alone is responsible for 40% of the Brazilian GDP, making it the richest state of Brazil.

Despite this, in the World Economic Forum Global Competitiveness Index (GCI), Brazil ranked 75th, declining from rank 57 in 2014/2015 (Schwab 2015). Brazilian transport infrastructure is inadequate. The country counts 214,000 km of roads (1600 km of them unpaved), only 30,000 km of railways, and 14,000 km of waterways. Around 60% of the cargo is being transported by road, even over long distances involving thousands of kilometers of travel. Thus, the costs of transport are very high (Mello 2012). In addition to poor infrastructure, supply chain managers face bureaucracy and complex customs regulations in Brazil. One project to improve the situation is the International Goods in Transit (TIM) initiative for Latin America and the Caribbean (LAC), sponsored by the Inter-American Development Bank. This initiative is introducing an electronic system for dealing with goods in transit.[1]

2.1.1.2 Russia

With the largest national landmass in the world, Russia covers about 17.1 million square kilometers (6,592,735 square miles) and nine time zones. The different parts of the country include areas of permafrost in Siberia and the Far North as well as taigas and steppes. Much of Russia's northern and eastern coastline is hemmed in by ice for much of the year, while its far eastern coast provides year-round access to major shipping ports along the Sea of Japan and the East Sea.

After the collapse of the Soviet Union in 1991, the country had to cope with a drastic decline in industrial and agricultural production. During the 1998 Russian financial crisis, the ruble experienced a depreciation to one-quarter of its former value. In its wake followed bankruptcies of major Russian private financial institutes and disappearance of large amounts of private and business assets. But with the increase in the price of crude oil, the Russian economy recovered.

The Federation profited from European demand in natural gas and oil and was able to generate an average GDP growth of about 7% annually in the period 1999–2007. In 2008, the growth rate was still at 5.2%, but Russia also felt the impact of the GFC. Since then, Russia has shown growth of an average of about 4% until 2013. In 2013, GDP growth reached 1.3 and 0.6% in 2014, due to the decline in demand for its commodities from its neighbors in Europe and weak domestic

---

[1]AEO stands for Approved Economic Operator. It is a standard term assigned by the World Customs Organization for supply chain security programs. AEO programs are structured within the WCO SAFE Framework of Standards to Secure and Facilitate Global Trade. The AEO program has been developing steadily since its inception and there are now 53 countries with operative AEO programs with over 30,000 certified companies. Additionally, there are another 11 country AEO programs in development. As of March 2014, 168 out of 179 World Customs Organization (WCO) members have signed letters of intent committing to implement the SAFE Framework. (Source: WCO, goaeo.com)

consumption (World Bank 2016a, b, c, d, e, f, g). In the GCI 2015–2016, the Russian Federation ranked 45th, improving from 53rd in 2014–2015.

Russia is an important export nation. The Russian outbound supply chain is strongly geared towards the transport of raw materials and natural resources, while inbound supply chains are oriented towards the movement of consumer and industrial goods. The Russian Federation's total value of exports reached $523 billion in 2013, making it number 9 among the world's top 10 export nations. Following Russia's annexation of the Crimea in 2014, and the multilateral trade restrictions, Russian total export values declined to $498.76 billion (Statista 2016).

Infrastructure is not well developed in most parts of Russia. In the West, which is more densely populated and features the majority of economic activity, the infrastructure is fairly mature, with additional infrastructural projects in the pipeline. Of 325 planned or ongoing infrastructure projects in Russia, about two-thirds are aimed at the Western part of the country. According to a survey by professional service company Ernst & Young (EY), 44% of these projects assume the participation of private investors, either public-private partnership (PPP) or fully private (Ernst & Young 2014).

In the East, the supply chain relies largely on railway transport. This, however, does not equate to Western standards. Russia has laid aside a significant investment budget for railway projects. This includes the high-speed railway program through 2030. According to the Ernst & Young survey, Moscow plans to invest $81.8 billion into railway transport. PPP projects are expected to spend another $380.6 billion in this sector. For roads and bridges, 77 projects are planned. At $162.1 billion, the largest part of Moscow's investment will go into the power and electric utility segment, including power supply, water supply, and gas supply infrastructure projects. All in all, Ernst & Young lists 148 projects planned through 2030.

Dubai-based port operator *DP World* and the Russian Direct Investment Fund (RDIF) agreed in early 2016 to form a joint venture to develop Russian seaports, transportation, and logistics terminals. DP World will invest $2 billion to upgrade port infrastructure throughout Russia (rt.com 2016). Since 2004, a network of federal highways, dubbed the Trans-Siberian Highway or AH6, spans the width of Russia from the Baltic Sea to the Sea of Japan in the Pacific Ocean. However, not all of these seven highways are well connected, especially in the central parts of Russia.

Russia is also thinking about better links with the US *Russian Railways* President Vladimir Yakunin and has proposed a plan for a massive trans-Siberian highway that would link the country's eastern border with the US state of Alaska, crossing the narrow stretch of the Bering Sea that separates Asia and North America. The Trans-Eurasian Belt Development (TEPR) would be constructed alongside the existing Trans-Siberian Railway, along with a new train network and oil and gas pipelines (Thompson 2015). Moscow would like to establish the country as hub for transport on the Asia-Europe-routes as well as cargo traffic between Northern Europe and India.

CSCOs who have to manage supply chains crossing the Federation frequently deal with challenges resulting from poor infrastructure and complex customs clearance processes and extreme weather conditions that make certain modes of transport impossible at different times of the year. Complex bureaucracy has not only limited economic growth in Russia but also the performance of the supply chain. Official approvals of important projects often drag along, sometimes discouraging investors. A more robust, seamless, and fluid supply chain would increase competitiveness in the global arena for exports and imports and would help Russia's domestic market too.

2.1.1.3 India

In 2014, India elected a new Prime Minister, Narendra Modi, who promised investments in infrastructure and job creation. During his period as Chief Minister of Gujarat from 2001 to 2014, the state grew richer and enjoyed faster GDP growth (10% annually) than India as a whole (Abad 2015). India's largest trading partners are China, the USA, and the United Arab Emirates. The top exports of India are refined petroleum, packaged medications, jewelry, rice, and cars, while its top imports include crude oil, gold, coal briquettes, petroleum gas, and diamonds (The Observatory of Economic Complexity 2016).

India's GDP growth rate reached 7.4% in 2015 (World Bank 2016a, b, c, d, e, f, g). A study by Harvard's Centre for International Development (CID) projected that India has the potential to be the world's fastest-growing economy until 2024, far outpacing China, which could slow to around 4.3% growth per year (Deshpande 2015). In the GCI 2015–2016, India ranked 55th, improving its position from rank 71 in 2014–2015.

India has the second fastest-growing service sector in the world, with an annual growth rate above 9% since 2001, which contributed to 57% of GDP in 2012–2013 (Bhargava 2015). During the same period, the industrial sector accounted for 25.8% of India's GDP. The agricultural sector contributed 17.4%. India is the world's second largest producer of agricultural products, accounting for 7.68% of the world's total agricultural output. Measured in GDP, India's industry sector ranks 12th in the world, and its service sector 11th.

Road is the dominant mode of transportation. Coastal, pipeline, and air transportation have been poorly supported for decades, and India's share of cargo transported by rail has declined steadily from over 85% in the 1950s to around 30% in 2015. The Modi government plans to change the trend through heavy investments in rail modernization. New logistics services are also emerging in India. The modern warehousing industry has seen noticeable improvement in the quality of warehouse facilities. Freight forwarding is moving increasingly towards building third- and fourth-party logistics (3PL/4PL) capabilities (KPMG 2010).

Despite the improvements, India is far from meeting international standards. Take India's 12 major harbors, for example: they are the most important interfaces for the global supply chain and hubs into the country. According to the Ministry of Shipping, maritime transportation covers around 95% of India's trading by volume and 70% by value. So far, however, the transport infrastructure to and from the harbors is insufficient. To help address this, while also providing a boost to the country's economy, the Indian Ministry of Shipping, Road Transport and Highways announced a massive investment in India's port and road sector. The government of India is planning to develop smaller low-cost seaports along the coastline to facilitate a more rapid movement of cargo (Indian Brand Equity Foundation 2015).

Beside the development of India's coastal and island infrastructure, Indian policymakers have also given thought and importance to international maritime stability. Prime Minister Narendra Modi said in February 2016, prior to the first Global Maritime Summit (Deccan 2016):

> A peaceful and stable maritime environment is critical for the regional and global security. It is also a must to harvest the riches of the oceanic ecosystems. Given the scale and complexity of the challenges, the international maritime stability cannot be the preserve of a single nation. It has to be a shared goal and responsibility of all the seafaring countries.

2.1.1.4 China

China has become an important trading partner to many countries. In 2013, China surpassed the USA and became the world's biggest trading nation measured by the value of imported and exported goods. In the GCI 2015–2016, China ranked 28th. China not only is the biggest consumer of energy, but it also has the world's biggest new car market and has the largest foreign currency reserves. A significant portion of China's trade involves importing raw materials and parts to be assembled into finished products and reexported. Rising wages and production costs in China, however, have made certain industries shift production to markets with lower labor and manufacturing costs, such as South Asia, Southeast Asia, and Africa.

The Chinese semiconductor market is still on the rise. Between 2003 and 2013, China's integrated circuit design sector revenues grew 24 times larger (from \$541 million to \$13,150 million). China's semiconductor consumption growth continued to exceed the average growth of the global semiconductor market: while the global semiconductor market grew 4.8% in 2013, China's semiconductor consumption grew by 10.1% and accounted for 55.6% of the global semiconductor market (PricewaterhouseCoopers 2015).

China has gradually raised the level of knowledge and skills to establish the production of higher-value products in country, leading to higher income and an increase in demand. By doing so, China has strengthened its own economy and raised the standard of living of its citizens. But foreign brands still play an important role. For example, international auto manufacturers are required to expand local production, generate employment, and support knowledge transfer. In August 2014, German auto parts suppliers operating in China were informed by Chinese authorities of the need to establish local partnerships, limiting their ability to operate independently (Taylor and Shen 2014). Such measures have been accompanied by additional trade duties, such as a customs tax of 25% on imported cars. By contrast, auto parts and production components are only liable to a 10–13% customs tax.

In 2014, total vehicle production in China reached 23.72 million, accounting for 26% of global automotive production (Wikipedia 2016a). Of the domestically produced passenger cars, two-thirds were manufactured by joint ventures. The largest of these automotive joint ventures is Shanghai's *SAIC*, a collaboration with *GM* and *Volkswagen* (Bilby 2014). In June 2012, *Shanghai Automotive Industries Corporation USA, Inc. (SAIC USA)* unveiled a North American Operations Center in Birmingham, Michigan. The objective was described as facilitating sourcing and procurement of automotive parts, materials, and components and working with almost 150 different North American suppliers (SAIC USA Inc. 2012). Certain parts for SAIC China are still manufactured in Germany. Manufacturing the unique parts in China would require specific skills and tools. Containers are driven to the port of Bremen by tractor-trailer and loaded onto barges destined for the container port at Bremerhaven. From there the parts travel to China, where, 76 days later, they are ready to be assembled (€uro 2013).

In 2015, *Daimler AG* paid $869 million for a 12% stake in its Chinese partner *Beijing Automotive Group (BAIC)*. Daimler secured two seats on BAIC's board, while the Chinese group increased its stake in their jointly owned passenger car manufacturer to 51%. This reshuffle should boost BAIC's valuation ahead of its planned IPO (Financial Post 2013). German cargo specialist *BLG*, based in Bremen, ships approximately 800 containers of parts from the port of Bremen to Daimler facilities all over the world, including India, South Africa, and China.

Nonetheless, local production ventures are not enough to spread meaningful economic opportunity beyond China's eastern coast, where most of the consumption, investment, and production are concentrated. China's interior provinces have not profited very much from the heightened levels of industrialization in the East and coastal regions of the country. In order to attract more investment in the interior and West of the country, the government has lowered corporate income taxes, from 25% of income to 15% in Chongqing, Sichuan, Guizhou, Yunnan, Tibet, Shaanxi, Gansu, Ningxia, Xinjiang, Inner Mongolia, Guangxi, and Qinghai. In order to raise the attractiveness of these regions, the People's Republic of China invested $118.7 billion in regional infrastructure such as railways, highways, and water and energy projects in 2015 (Chinadaily.com 2015).

China is moving away from investment-driven exports towards more service-oriented models based on increased domestic consumption. China's service sector accounts for about 46% of GDP. In 2013, services surpassed industries, such as construction and manufacturing, for the first time. The service sector includes transport, storage, and post (5% of GDP), wholesale and retail trades (10%), hotel and catering services (2%), financial services (6%), real estate (6%), and a mix of services categorized as "other" (18%) (Bajpai 2014).

The more robust transport connections to the hinterland are creating new logistics demand. The development might lead to a diminishing importance of certain ports, with a shift from southern regions to northeastern regions. New challenges for the supply chain will arise, since the critical mass for cost-effective transportation and customs clearance has not been reached.

With various tolls levied by provincial and city governments and many areas of inefficiency, logistics cost, with variations by region, amount to 18% of China's GDP, higher than in many developed countries. KPMG estimates, in their paper

"On the move in China," that for trucking companies, the tolls can account for one-third of total costs (KPMG 2011). Other reasons for the high costs are fragmentation of the market, the lack of modern logistics expertise across the country, and a lack of standardization. In order to improve the situation in the logistics industry, the government established a 5-year plan in 2014. The goal is to decrease the overall logistics costs by 2% points.

China has become the world's largest e-commerce market, with revenue estimates as high as 4 trillion yuan ($62 billion) for 2015, up 33.3% from 2014. Consequently, express deliveries have risen to 20.6 billion in 2015 (Yinan 2016). This growth requires significant investment in the transport and logistics industry.

Jack Ma, founder of Alibaba, a business and consumer online trading platform, invested $1.5 billion to build a nationwide network of warehouses (Erickson 2011). Alibaba Group's Taobao subsidiary, China's largest online retail website, accounts for nearly 70% of all C2C and B2C transactions. Taobao formed partnerships with national and international logistics and delivery companies (A.T. Kearney 2011).

2.1.1.5 South Africa

South Africa experienced an average growth rate of approximately 5% between 2004 and 2007. The period from 2008 to 2012 recorded average growth of 2.2%, while for 2014, South Africa recorded a 1.5% and for 2015 a 1.3% growth in GDP. One of the reasons for the slowdown in growth is the South African government's high levels of foreign debts, which in 2014 totaled $136.6 billion or 38.2% of the country's GDP. According to *Forbes Magazine*, $60.6 billion of South Africa's external debt is in foreign currencies, a significant burden considering that the South African Rand declined 15% against the US dollar in 2014 compared to 2013 (Colombo 2014).

Among the country's most important trading partners are the EU, China, Japan, and the USA. South Africa's primary exports to the EU are fuels and mineral products, machinery and transport equipment, and other semimanufactured goods. The EU and South Africa have established a Trade, Development and Cooperation Agreement that covers 90% of bilateral trade. The second largest industrial sector in South Africa is manufacturing, ranging from textiles and metals to electronics and automotive. BMW and VW have established plants in the region (BMW 2016).

Trade with the USA is helped along with the preferential trade pact known as the African Growth and Opportunity Act (AGOA), which gives South Africa duty-free access for exports to the USA worth as much as $1.7 billion a year. AGOA has enabled South Africa to more than double its exports to the USA since 2000 (Vollgraaff 2015). In December 2015, the heads of state of South Africa and China signed a number of trade and industry investment agreements, including a memorandum of understanding on joint development of the Silk Road Economic Belt and the 21st Century Maritime Silk Road – the Belt and Road (Saoutafrica.info 2015). In the GCI 2015–2016, South Africa ranked 49th.

South Africa is a logistics hub. When it comes to the supply chain, South Africa is considered one of the best among the BRICS nations. The Logistics Performance Index 2014, published by the World Bank, ranked South Africa 34th out of

160 countries assessed, placing the country second of the Big Four after China, ranked 28th (Brazil was 65th; Russia, 90th; and India, 54th) (World Bank 2014). In 2016, South Africa climbed to 20th (World Bank 2016a, b, c, d, e, f, g).

However, when considering important aspects not measured by the World Bank, including the cost of domestic transport and logistics, South Africa does not perform as well. One of the reasons is that road density in South Africa is low and the cost of exporting goods is higher than $2000 per container, which is more than twice as much as in other BRICS nations. Also, the lead time required to export via road is 30 days, compared to 13 days in other developing countries. Poor skills and low technology utilization are hindrances in improving supply chain efficiencies in South Africa (KPMG 2013).

2.1.2 CIVETS

South Africa, along with Colombia, Indonesia, Vietnam, Egypt, and Turkey, forms the so-called CIVETS states, the next group of emerging markets expected to outperform economically over the next decade. South Africa was already discussed in the BRICS section.

2.1.2.1 Colombia

Colombia is the third largest economy in Central and South America. While the USA is Colombia's largest trading partner, Columbia ranks only 19th among US trading partners. The U.S-Colombia Trade Promotion Agreement, which took effect on May 15, 2012, aims to remove barriers to expanding trade by eliminating tariffs on 80% of U.S consumer and industrial goods. US exports to Colombia include oil, machinery, agricultural products, and organic chemicals. U.S imports from Colombia include crude oil, gold, coffee, and cut flowers (Villarreal 2014). Colombia's GDP growth rate averaged just above 1% from 2001 until 2015, mostly due to the rise in commodity exports. In order to reach its full growth potential, Colombia needs to improve infrastructure and security and reduce income inequality (Trading Economics 2016b). In the GCI 2015–2016, Colombia ranked 61st.

Improving the logistics performance will have an impact on the economic growth. On the Logistics Performance Index 2016, Colombia came in 94th, slightly improving from 97th in 2014. Policymakers are aware of the fact that logistics performance has to be improved and cost brought down. Spending in the transport sector has already increased. It accounted for 2% of GDP between 2009 and 2013, compared to 1% between 2002 and 2008 (International Monetary Fund 2015).

2.1.2.2 Indonesia

The Indonesian economy benefits from strong domestic consumption. The World Bank states that Indonesia is now one of Asia Pacific's most vibrant democracies, which maintains political stability well and has emerged as a confident middle-income country. In 2014, GDP grew at a rate of 5%. The country's gross national income per capita has steadily risen, from $560 in the year 2000 to $3650 in 2014.

Indonesia is the world's fourth most populous nation and the world's tenth largest economy in terms of purchasing power parity and one of the leaders of the ASEAN and the Asia-Pacific Economic Cooperation (APEC) (World Bank 2016a, b, c, d, e, f, g).

In 2008, Indonesia became a member of the G20.[2] It is the only ASEAN country to be included so far. With the USA, Indonesia has signed a Comprehensive Partnership in 2010, covering technical assistance, capacity building, technology, and ideas that foster innovation and reform.

In 2014, Indonesia shipped $176.3 billion worth of goods around the globe, an increase of 11.7% since 2010. Its primary exports are coal briquettes, palm oil, petroleum gas, crude petroleum, and rubber (Simoes 2016). The main export destinations are Japan, China, the USA, Singapore, and India. The number one exported good to the USA is clothing (The World's richest countries 2016). Recently, China has become a major trading partner to Indonesia. In 2014, China's imports from and exports to Indonesia exceeded the USA's. China is a party to a regional FTA that includes Indonesia and is negotiating the Regional Comprehensive Economic Partnership (RCEP) with Indonesia and 14 other countries. In the GCI 2015–2016, Indonesia ranked 37th.

In reference to the country's logistics performance in 2016, Indonesia ranked 63rd, sliding slightly from 53rd in 2014. Poor infrastructure in the country leads to congested roads. Many of the ports are in poor condition. According to the OECD Economic Survey Indonesia 2015, the government made maritime connectivity one of the major policy focuses. The goal is to enhance remote-island links and upgrade port infrastructure. The initiatives include the establishment of ten deep-water ports (OECD 2015).

2.1.2.3 Vietnam

As a member of the ASEAN and APEC regional blocs and potential TPP member, Vietnam's economic boom can be traced to its shift from a centrally planned economy to a market-led, international trade-driven country in 1986. In 2014, Vietnam recorded a GDP growth rate of 6%.

According to the World Bank, Vietnam improved macroeconomic stability, greatly reduced poverty, and has achieved universal primary education (World Bank 2016a, b, c, d, e, f, g). Education and training is vital for logistics, because the industry requires skilled workers. Vocational training is lagging. Currently, nearly 300,000 businesses in Vietnam provide logistics services, 40% of which are located in Ho Chi Minh City (HCMC). HCMC University of Transport and the Vietnam Maritime University have opened new facilities that will focus on logistics education. Vietnam's manufacturing sector has been a strong driver for growth in

---

[2]G20: The Group of Twenty (G20) includes 19 individual countries – Argentina, Australia, Brazil, Canada, China, France, Germany, India, Indonesia, Italy, Japan, South Korea, Mexico, Russia, Saudi Arabia, South Africa, Turkey, the UK, and the USA – along with the European Union (EU), which is represented by the European Commission and by the European Central Bank. These economies combined command 85% of the global economy.

transport and logistics services. The total logistics-related workforce in Vietnam is estimated at 1.5 million people. In the LPI 2016, Vietnam ranked 64, sliding from position 48 in 2014. In the GCI 2015–2016, Vietnam ranked 56.

Ports are important for the country's logistics sector, as much of Vietnam's commercial and manufacturing activity is concentrated along the Mekong River Delta and in HCMC. In 2014, volume handled at Cai Mep, HCMC's deep-water port increased by 16.7% year over year. At the northern port of Hai Phong, volume increased by 14.3% during the same period. Analysts expect the region's seaports, which handled 200 million tons in 2015, to handle around 650 million tons in 2030 (Laursen 2015). Airfreight volumes have grown at an annual rate of 15.3% in Vietnam, and experts predict an average annual growth rate of 11.5% over the next 10 years.

2.1.2.4 Egypt

Although among the least politically stable CIVETS states, Egypt is the second most industrialized country in the group, after South Africa. In 2015, Egyptian nonpetroleum export revenues dropped by 19% year-on-year (Alsahary 2015), while tourism revenues declined by 13% during the same period (Kortam 2015). The business hubs are Cairo and Alexandria, as well as the Canal governorates (Ismailia, Port Said, and Suez). However, the Egyptian government is taking steps to integrate the Nile Delta and Upper Egypt by promoting investment to alleviate poverty (African Economic Outlook 2015). In the LPI 2016, Egypt ranked 49th, improving from 62nd in 2014. In the GCI 2015–2016, Egypt ranked 116th.

2.1.2.5 Turkey

Turkey is a geopolitical hotspot, with favorable demographics and a diversified economy. The country is positioned as a strategic bridge between the Middle East and Europe – economically, culturally, and logistically. However, Turkey is economically exposed to the situation in Ukraine and sanctions against Russia, as well as the ongoing conflicts in Syria and Iraq. Nevertheless, the Turkish economy grew 3% in 2014 (World Bank 2016a, b, c, d, e, f, g). In 2014, Turkey's exports reached an all-time high, hitting $157.6 billion, reflecting a 4% year-over-year increase. Turkey's single largest trading partner is the EU-28, which account for about 43% of Turkey's exports in 2014 (The Republic of Turkey Prime Ministry, Investment Support and Promotion Agency 2016).

The situation in Turkey has deteriorated in 2016, with total exports registering a 14.4% decrease in January 2016, when compared to January 2015 (Exporters' Assembly of Turkey 2016). Turkish exports to Germany, the USA, and Russia dropped by 1.5%, 6.6%, and almost 30%, respectively. In the current environment, Saudi Arabia emerged as an increasingly important trading partner for Turkey, with exports to the Kingdom increasing by 30.5%, followed by a 15% increase to Egypt, and a 9% increase to Israel (Hurriyet Daily News 2016).

The negotiations regarding the TTIP are important for Turkey, because its exclusion from the pact could cost the Turkish $20 billion per year. According to a study by the US-based Brookings Institution and the Turkish Industry and

Business Association (TÜSİAD), the exclusion of Turkey from the TTIP would cause a significant trade diversion in Turkey's trade alliance and cause an increase in the country's $8.5 billion (2012) trade deficit with the USA. The reason for Turkey's concerns being that "the goods of any third party that the EU has a free trade deal with can enter Turkey with zero duties, but the decision to provide the same privileges to Turkey is up to the third party" (Hurriyet Daily News 2013).

Turkey has been in negotiations with the EU for years. The Turkish government signed a customs union agreement with the EU in December 1995. Turkey has announced the possibility of freezing this agreement in the event that Turkey does not become a party to TTIP. Currently, Turkey is asking the EU to add an article into the TTIP for the automatic inclusion of countries with whom the Union already has a customs union arrangement in place (Zuvin and Kurtuldu 2015). In the LPI 2016, Turkey ranked 34th, sliding from position 30 in 2014. In the GCI 2015–2016, Turkey ranked 51st.

The situation in Syria takes a toll on the Turkish carriers that service the Middle East and have to cross Syria. The alternative would be for Turkish carriers to load the trucks onto ferries and have them shipped to Lebanon, Egypt, Israel, or Saudi Arabia. However, this increases travel time and costs despite subsidies from the Turkish government (Kauffmann Bossart 2015).

Traffic and trade between the Turkish province of Hatay and Syria has been troubled. In 2010, approximately 110,000 Turkish trucks shipped fruits, vegetables, and building material to Syria. In 2013, the number of trucks crossing the border had declined to 12,200. Additionally, trucks don't haul loads into the inland, but merely move within a 1.5 km-wide buffer area beyond the Syrian border. Here, the cargo is taken over for distribution by a subsidiary of Al-Qaeda, according to one Swiss newspaper (Kauffmann Bossart 2015).

Despite these challenges, Turkey remains determined to increase international trade. Turkey is part of the MIKTA, an innovative partnership between Mexico, Indonesia, South Korea, Turkey, and Australia. The platform was inaugurated in 2013 and aims at "advancing the common interests of the international community." For example, MIKTA has published joint statements on various global issues from climate change to North Korean nuclear tests (Australian Government 2016).

2.1.3 MINT

BRICS countries' economic growth has slowed. So, investors have turned their attention to Mexico, Indonesia, Nigeria, and Turkey (dubbed MINTs), grouped together because of the large populations, favorable demographics, and economic relevance and growth. Two of the MINT countries, Indonesia and Turkey, have already been covered under CIVETS.

2.1.3.1 Mexico
Mexico has signed 12 FTAs with 45 countries, among them the EU and Japan. Being part of NAFTA has had a deep impact on Mexico's economy and positioned

it as one of the top three trading partners of the USA. Conversely, the USA is Mexico's single most important trading partner. In the wake of rising wages in China, Mexican businesses are hoping for more nearshoring of business from US companies.

Mexico is diversifying its foreign trade structure and has approached partners in the Latin American region. In 2011, Peru, Chile, Colombia, and Mexico formed the Pacific Alliance, aiming at free trade and economic integration. The four founding nations of the Pacific Alliance represent nearly 36% of Latin American GDP; combined, they represent the sixth largest economy in the world with a GDP (measured by purchasing power parity: PPP) of more than $3 trillion (Wikipedia 2016b). In the LPI 2016, Mexico ranked 54th, sliding slightly from position 50 in 2014. In the GCI 2015–2016, Mexico ranked 57th.

2.1.3.2 Nigeria

Oil is Nigeria's main source of foreign exchange earnings and state financing. Following the 2015 global decline in oil prices, the Nigerian economy has suffered. Since the domestic security situation also presents a challenge, experts assume significant downside risks to the economic, commercial, and financial landscape in Nigeria (PwC Economics & Policy 2015).

The African Development Bank Group, however, is counting on the non-oil sector in Nigeria to make up for the losses caused by declining oil prices. According to the Bank, the service sector will be a key driver for non-oil growth in Nigeria and already accounts for 57% of GDP. Manufacturing and agriculture, respectively, contribute about 9 and 21% to the Nigerian economy, while growth in retail and wholesale trade, real estate, information, and communication are helping to diversify the economy (African Development Bank Group 2016). In the LPI 2016, Nigeria ranked 90th, sliding slightly from position 75 in 2014. In the GCI 2015–2016, Nigeria ranked 124th.

While importing auto parts through Durban in South Africa takes 1 month, importing the same parts through Port Apapa, Nigeria, will take over 3 months. This is mainly due to port-capacity limitations and differences in handling times (Avasthy et al. 2015).

2.2   Transformational Growth

The experience in the emerging markets shows that investments in infrastructure, skills, and connectivity, combined with political stability and fiscal incentives, create a virtuous cycle of increased economic activity. Drawn by lower cost in wages, energy, land, etc., companies and industries migrate to other markets, from the West to the East in the past decades and from China to Southeast Asia more recently. The process will probably continue on and on, with countries moving up or down in attractiveness for investors, traders, and manufacturers.

One of the primary reasons companies migrate to other countries is the reduction of competitiveness and better conditions elsewhere. This is usually due to uncompetitive or even prohibitive costs of operation and unfavorable regulation, as well as constraints in infrastructure and available talent. In order to avoid relocation, companies first reduce costs by rationalization and cutting back jobs. If such measures are not successful, management moves production to lower cost locations. Another reason for moving is closeness to markets the company wishes to serve. The digital revolution, in particular the Internet of Things, makes decentralization and moving factories to consumer markets possible through easier quality control and risk management.

Transfer of industries can also result from public pressure. On March 11, 2011, an earthquake-generated tsunami off the coast of Japan caused three nuclear meltdowns at the nuclear plant in Fukushima, Japan. The following year, 100,000 antinuclear protesters marched through central Tokyo to voice their opposition to atomic power. As a result, all 50 of Japan's nuclear reactors were taken offline. The global financial crisis has shown that governments need to diversify the economy to raise resilience against external shocks. Therefore, countries such as China are transforming from a predominantly export-led economy towards one driven by stronger domestic demand.

Diversity and density of supplier networks, available labor, and infrastructure are important drivers of economic success. This is demonstrated by the example of the manufacturing of electronic devices such as the Apple iPhone. Until a decade ago, Apple manufactured its products in the USA. Today, a major part of the world's components is manufactured in China, due to its robust network of manufacturers and suppliers. Assembling devices anywhere else, where manufacturing networks are less developed, would create logistical challenges, increase production cost, and reduce flexibility. Experts state that it would cost about $65 more to assemble the iPhone in the USA than in China, where the costs are estimated at $8 (Blodget 2012). Governments are able to stop companies from relocating through tailored measures. This includes education and training programs, immigration policies for required talent, and trade facilitation measures such as the optimization of cross-border ecosystems by improving market access, administrative processes, logistics services, digital and transportation infrastructure, business environment, and security.

Economic development, competitiveness, and prosperity of nations are closely related to creativity, as referenced by the Global Creativity Index (GCI) from the Martin Prosperity Institute at the University of Toronto's Rotman School of Management. The GCI specifically states that advanced economic growth and sustainable prosperity are based on the three Ts of economic development – talent, technology, and tolerance. The authors of the GCI 2015 emphasize a new creative capitalism on the rise. The index is measured by factors such as investments in R&D, the share of adults in higher education, and the individually felt degree of tolerance of people living in cities. In short, creative, tolerant, and technologically advanced nations will be among the winning players of tomorrow. As reported by the GCI 2015, those nations are Australia, which takes the top spot, the USA (second place), New Zealand, Canada, and Denmark and Finland, both in fifth place. Sweden, Iceland, Singapore, and the Netherlands round out the top ten out of 135 nations analyzed (Florida et al. 2015).

Supply and value chains need to be competitive. A detailed supply and value chain mapping will identify at which level certain business sectors need what kind of support. By the same token, policymakers will have to support the launch and growth of innovative industrial sectors in order for the industry to stay competitive and create economic activity and jobs over the long term. Governments need also to ensure that the infrastructure and logistics capabilities support the growth scenario.

Holding on to uncompetitive sectors binds resources, which are better used for transition towards more promising industries. In the long run, a society does not benefit from preventing the shift of industries that are not competitive. When protective measures are imposed such as punitive tariffs, consumers and taxpayers pay the price. The former will face more expensive goods and services; the latter subsidize uncompetitive businesses. Entrepreneurship and innovation suffer, while foreign companies are less likely to invest. Instead, a more strategic management of long-term inevitabilities is required. Economies must be transformed by actively transitioning from waning industries, constantly repositioning established industries, and creating the necessary policy, economic, and infrastructural foundations for future industries to rise.

As more cost-efficient global talent pools and new markets come online, new players, introducing new technologies, will emerge from unexpected places, forming new value chains and launching new business models. Not only companies and industries but also economic blocs, regions, countries, and even cities compete for talent and investment, honing particular advantages and convening new ecosystems. The flexibility required to be and stay competitive requires access to a largely unobstructed exchange of goods and people, data, and information. Free exchange itself can act as a driver of change and transformation. New players and new knowledge result in new value chains, which can provide new routes for innovation, increased efficiency, and competitiveness. Conversely, in closed and isolated markets, whole industries and economies can lapse as a result of obstructed information, investment, talent, or trade flows. This can be illustrated by looking at the stunted economic development of ex-Soviet satellite and client states over the past several decades.

Entrepreneurship, expertise, technology, and the supply chain itself are key drivers of transformation. In order to benefit from investments in the long run, countries should specifically promote business activities in those industries that are in a position to control most parts of the supply and value chain. In this respect, it is not necessary that the country possess most of the resources. No nation possesses all the resources and knowledge within its borders to sustain a modern economy. International relations and global integration are prerequisites for long-term prosperity and well-being. The economic success of a nation is a function of the performance of the supply chain, with economic surpluses resulting from the efficient operation of vertical supply chains, tailored for specific industries.

One good example is Switzerland: the country has neither cocoa nor sugar nor a large work force. Nevertheless, it is one of the most renowned, successful producers of chocolate in the world. This results from excellent supply and value chain management, in addition to building and marketing brands for global distribution. Switzerland ranked number 1 in the GCI 2015–2016 and 2014–2015. Therefore, it is fair to assume that the real driver of success in trade and business comes down to the health and depth of related supply and value chains, the control of the flow of goods along the various steps towards a finished product, and its marketing and distribution.

Also important for economic stability and growth is the small business sector. In *Forbes*, Kelly Phillips Erb wrote:

> small businesses, like taco trucks, do have a significant impact on our communities and in our cities. According to the Small Business Administration (SBA)[3], 28 million small businesses in America account for 54% of all U.S. sales. Those same small businesses provide 55% of all jobs and have provided 66% of all net new jobs since the 1970s (Erb Phillips 2016).

Only attracting businesses is not enough. Businesses must constantly repeat the assessment of benefits for long-term planning and viability. In the case of diminishing conditions, businesses will consider shifting to new markets and bases of production. Therefore, the transformation of economies needs to be a continuous effort.

Health of people and the planet should be key considerations too and not be second to the economic thinking and acting. The key to sustainable prosperity and well-being of nations is the collective balancing of increased efficiency and competitiveness with an acceptable standard and quality of living. What the emerging countries are experiencing today in terms of environmental pollution developed countries faced during the First Industrial Revolution. The magnitude, however, is very different. Policymakers in developed and developing countries are grappling with the effects of environmental degradation. Governments all over the world are searching for solutions to protect against everything from more frequent and extreme weather events to the rapid spread of global pandemics. Beside local, national, and regional measures, more collective and integrated global action will be needed in future.

A positive step in this direction was taken at the annual UN Climate Change Conference in Paris. The aim of the Paris conference was to establish subsequent agreements following the expired Kyoto Protocol, which was, up to now, the only internationally binding instrument regarding climate policy. At the annual meeting in Paris in 2015, the 195 nations agreed to "pursue efforts" to limit the global temperature increase to 1.5 °C. Nonetheless, stronger commitments (and enforcement mechanisms) will be required to achieve the stated objective going forward.

---

[3]To be found at https://www.sba.gov/managing-business/running-business/energy-efficiency/sustainable-business-practices/small-business-trends

2.2.1   Country Strategies

Countries select different strategies and focus on different industries and geographies to ensure future competitiveness. Since 1978, for example, the Chinese economy has been in a state of continuous transformation. Once a state-dominated economy, the country has become the world's largest exporter with many of its industries and private companies completely integrated into global supply and value chains. More recently, though, as a result of steadily increasing wages and higher energy and land costs, China has faced the growing shift of production towards lower cost destinations in Africa, South Asia, and Southeast Asia. China is therefore looking for new ways to remain competitive.

One way is to replace human capital with robots. Automation is being promoted in various industrial sectors. This is also intended to cover labor shortages in the automotive, metal processing, and even food and beverage industry. Despite the enormous population, officials fear labor shortages of as high as 30% in some areas (Shih 2015). The global robotics industry is expected to grow from $28.3 billion in 2015 to $151.7 billion by 2020, according to a report from Tractica, a market intelligence firm that focuses on human interaction with technology, and China's market share has increased, up from 25% in 2013 to 31% in 2015, according to the International Federation of Robotics (IFR) (Mullen 2016).

As a consequence of rising costs, production has also been shifting from the more expensive East of the country towards the less expensive Western regions. This move has been heavily promoted by the Chinese government, but the necessary active supply chain ecosystem is largely missing in the Western regions of the country. Various service providers offer transport of complete loads and containers; however, there is a lack of cost-efficient transport solutions for less-than-container or truckload shipments.

With its vision for greater logistical and economic integration in the West of the country and an industrial policy prioritizing the development of a homegrown robotics capability, the People's Republic of China aims at a smooth and future-oriented transition to the next stage of its development. In the northeastern province of Liaoning, a large industrial robotics capacity will be ready for use by the end of 2017. The Liaoning provincial government has promised tax reductions and venture capital support. The program will have an estimated annual output of $8 billion (Asian Scientist Magazine 2013). By strategically developing industrial robotics capacity, China hopes to retain its competitive edge in manufacturing while also becoming a major player in the global market for robotics technology. It remains to be seen whether or not Chinese policymakers will be as successful in enabling an end-to-end domestic robotics ecosystem and supply chain as they were in laying the groundwork for the domestic manufacturing of electronic devices.

Singapore, on the other hand, benefits from its central location in Southeast Asia and its status as a hub for global supply chains. The city-state is known for its quality of living, hyperconnectivity, innovativeness, and the leadership of its policymakers. The country's international airport offers direct flights to many world cities, with those in ASEAN reachable within 3 h or less. The country's innovative strategy relies on its role as a modern hub for trade between ASEAN, China, and the world.

Singapore's most enduring competitive advantages can be traced to a number of inspired and rather large infrastructure investments between 1963 and 1975. This has included construction of a deep-water terminal at the Port of Singapore, a doubling of the country's energy capacity, and the strategic interconnection of water resources with Malaysia. Today, the maritime industry contributes about 7% of the country's service-oriented GDP, and the country is home to three of the world's busiest container ports. Adding to the total package is a progressive free trade policy and relative political, economic, and legal stability. Singapore was ranked first for logistics services in Asia in a 2014 study by the World Bank. Singapore has also led on environmental policy and boasts a fully integrated waste system and value chain, from collection and recycling to waste treatment, energy recovery, and landfill management.

2.2.2 International Collaboration

The opportunities for policymakers to take part in the design of future economic models and supply and value chain ecosystems are diverse. Governments influence the layout of global supply chains, as well as behavior of other nations, by promoting free trade or applying protective measures. For policymakers, a general grasp of the economics of free trade and the basic mechanisms of global supply chains is vital. The consequences of trade-restrictive measures taken have to be realistically assessed.

Policymakers need to maintain a perspective on the entire global supply and value chain: How are markets connected to one another? What does this connectivity really mean? What are the strengths and weaknesses of a market and its main trading partners? What does a certain stage of economic development and technological maturity imply for the next phases of evolution? What effect do trade facilitation or trade barriers have on the economy and the ability to grow? How do the trade policies impact the relations with neighbors and trading partners? These are important considerations when formulating industrial and trade policy. Policymakers have to think about future ecosystems in the context of global supply and demand and drive the necessary transformation accordingly. Economic blocs have been formed under such considerations and significantly helped the development of nations.

One example of a successful economic bloc is the *Association of Southeast Asian Nations (ASEAN)*.[4] ASEAN was founded in 1967 and is one of the success stories of modern economics. In 2014, the region was the seventh largest economic power in the world (ASEAN 2015a, b). It was also the third largest economy in

---

[4]More information can be found at the Council on Foreign Relations: http://www.cfr.org/asia-and-pacific/asean-association-southeast-asian-nations/p18616

Asia, with a combined GDP of $2.6 trillion, which is higher than India. Between 2007 and 2014, ASEAN trade increased by a value of nearly $1 trillion. Most of that (24%) was trade within the region, followed by trade with China (14%), Europe (10%), Japan (9%), and the USA (8%). During the same period, foreign direct investment (FDI) in the ASEAN member countries rose from $85 billion to $136 billion.

In 2015, the ASEAN states established the *ASEAN Economic Community (AEC)* as a way forward to promote economic, political, social, and cultural cooperation across the region.[5] The idea is to move Southeast Asia towards a globally competitive single market and production base, with a free flow of goods, services, labor, investments, and capital across the ten member states. With 622 million people, ASEAN has the third largest labor force behind China and India.

The AEC's vision for the next 9 years, laid out in the AEC Blueprint 2025, calls for (ASEAN 2015a, b):

1. A highly integrated and cohesive economy
2. A competitive, innovative, and dynamic ASEAN
3. Enhanced connectivity and sectoral cooperation
4. A resilient, inclusive, people-oriented, and people-centered region
5. A global ASEAN

With a diversity of economic, industrial, and technological development among AEC member states, full alignment on the community's integration timetable has been a struggle.

The AEC must bolster intra-regional trade to reduce its vulnerability to external economic shocks. This requires a common regulatory framework to address infrastructure gaps and the simplification of administrative policies, regulations, and rules. Only 50% of ASEAN businesses have utilized tariff reductions set out in the ASEAN's regional free trade agreement (FTA). Although intra-regional tariffs are in decline, nontariff measures (UNCTAD 2016), health and safety regulations, licenses, and quotas are on the rise and need to be addressed. According to a study by the International Labour Organization (ILO) and Asian Development Bank, the AEC could generate 14 million new jobs and boost the region's economies by 7.1% per year between 2016 and 2025 (International Labour Organization 2014), which is more than ASEAN's annual growth of 5.4% from 2004 to 2014 (Halley and Torné 2015).

Many companies have already begun approaching ASEAN as one market. This has been helped by the ASEAN Single Window (ASW), a regional initiative to allow free movement of goods across borders. But progress remains relatively slow: the region can only proceed at the behest of national governments, and with every ASEAN country so different, a common vision can be hard to arrive at.

---

[5]Further information can be acquired at http://asean.org/asean-economic-community/

The AEC is not the only trade agreement in Southeast Asia. Three months before the kick-off of the ASEAN Community, four ASEAN countries (Brunei, Malaysia, Singapore, and Vietnam) signed up to the *Trans-Pacific Partnership (TPP)* with another five AEC countries (Cambodia, Indonesia, Laos, the Philippines, and Thailand) interested to join (Sim 2015). The TPP is a binding agreement, facilitating greater trade and economic integration between countries in Asia, North America, and Latin America. Countries with high export potential, such as Malaysia and Vietnam, are expected to benefit significantly from TPP. Those that did not sign up for the partnership risk losing out. This could have a disruptive effect on the ASEAN region.

Adding to the network of economic regions is China, the world's second largest economy and not part of the TPP. China is in the process of forming its own, parallel economic bloc: the *Regional Comprehensive Economic Partnership (RCEP)*. This alliance will comprise ASEAN, Australia, China, India, Japan, South Korea, and New Zealand.[6] Once all agreements are activated, Singapore and Vietnam would be the first two ASEAN countries with free trade access to Europe (via individual agreements), the USA (through TPP), and Asia (through the AEC and the RCEP).

ASEAN economies not currently included in the TPP might appreciate a more fluid supply chain reaching beyond their regional bloc, by way of those ASEAN members that have signed up for TPP. However, the exacting TPP standards could pose challenges for many countries in the region, due to stringent requirements on enhancing practices, meeting higher quality of production standards, and a raft of related rules and regulations. Despite the complexities of implementing new trade agreements, though, direct and indirect entry routes for AEC countries into the TPP and RCEP markets (which account for 40% and 30% of global GDP, respectively) will bring substantial economic benefits to all parties involved.

In a globally interconnected and interdependent world, national policymakers must wield their influence responsibly. The scale and scope of the global challenges of the twenty-first century call for a new level of collective action. National policies should be formulated with a focus on mitigating broader systemic risks to the future and sustainability of the entire global ecosystem.

2.3   Interdependence and Resilience

The connectedness and interdependence of today's world became evident in 2007/2008. The banking and property crisis in the USA, its climax being the bankruptcy of Lehman Brothers in 2008, triggered one of the most severe economic crises in global history. Volume of world goods exports dropped from a positive growth of 6.5% in 2007 to 2.3 in 2008 and −12.1% in 2009 (World Trade Organization 2012).

---

[6]Further information at https://www.mti.gov.sg/MTIInsights/SiteAssets/Pages/FACTSHEET-WHAT-YOU-NEED-TO-KNOW-ABOUT/Factsheet%20on%20RCEP%20(June%202014).pdf

Many countries suffered from the global financial crisis (GFC), which unleashed a well-known series of effects on the global market. The European monetary system came under severe pressure. The European Union's GDP fell by about 4% in 2009, its worst performance on record. In Russia, exports during the first half of 2009 were only about half of what they had been during the corresponding period in 2008. Africa, having a weak connection to global financial markets, felt the impact of the crisis in the drop of commodity prices. Although world trade bounced back by the third quarter of 2009, the growth stayed below expectation. The long-term implications of the GFC still remain to be seen. As Otaviano Canuto, the executive director of the IMF, wrote in the Huffington Post, "world trade suffered another disappointing year in 2015, experiencing a contraction in merchandise trade during the first half and only low growth during the second half" (Canuto 2016).

The main characteristics of the new era of global trade are relatively mature supply and value chains, increasing regionalization and localization of production, the rise of protectionism and nontariff barriers after a period of significant tariff reductions, and an e-commerce boom, which struggles to overcome the hurdles in international business. Annually, FM Global, a commercial and industrial insurance company, reports how unforeseen events affect the global supply chain. The "Resilience Index" provides a ranking of 130 countries and territories according to their business resilience to supply chain disruption. The scores that generate the ranking are calculated as an equally weighted composite of nine core drivers that affect resilience significantly and directly. Nine key drivers of resilience are aggregated into three broad factors, which combine to form the index. The factor economy includes GDP per capita, political risk, and oil intensity. The factor risk aggregates the indices of quality accumulates, exposure to natural hazard, quality of natural hazard risk management, and quality of fire risk management. Control of corruption, infrastructure, and local supplier quality form the factor supply chain. In 2015, Norway proved to be the country with the most resilient supply chain. The USA is divided into three regions based on states having a dominant natural hazard: region 1, wind; region 2, earthquakes; and region 3, miscellaneous. Region 3 is ranked tenth for their business resilience to supply chain disruption; regions 1 and 2 are ranked 16th and 21st (FM Global 2015).

The "FM Global Resilience Index" depicts how a bundle of factors impacts the supply chain. Regional disturbances of the aforementioned kind caused disruption in the past and had a global impact, such as the fire that broke out at one of the production facilities of the *Toyota*-subsidiary *Aisin Seiki Co.* in 1997. The plant had to close for 2 weeks. Due to in-time delivery, the Toyota car factory ran out of parts quickly. Sale losses were said to accumulate up to $325 million.

The year 2010 alone brought forth 960 natural disasters around the world, among them the eruption of the Icelandic Eyjafjallajökull volcano, which leads to airspace closure due to airborne ash in one of the major air transport routes in and around Europe for almost 2 weeks. The results included missing parts for automotive plants in Munich and a cell phone manufacturer in Korea and shortages of fresh fruit and fish, which usually come in by air. Estimates of the economic impact range from $1 to 3.3 billion (Demand Caster 2010). In 2011, tornadoes and floods in the USA caused $32.6 billion in direct insured losses, nearly double the $18.6 billion in catastrophe-caused direct insured losses that insurers generally incur over the first 9 months of any given year (Homeland Security News Wire 2012). In the same year, the worst Thai flood in 50 years hit supply chains of global players such as *Apple* and *Toyota*. The Thai government estimated damages of as much as 120 billion baht ($3.9 billion) (Yang 2011); some experts claim that this figure might be even higher, since the flooding disrupted more than 14,000 businesses in a country that produces about a quarter of the world's disk drives and serves as a production hub for Japanese car makers. These examples show the importance of a high degree of flexibility in the supply and value chain and the ability to act quickly when disruptions occur.

Literature

€uro 11/2013. *Dossier: So kommen deutsche Autos nach China* (p. 24 ff).

A.T. Kearney. (2011). *Ideas and insights, China's E-commerce market: The logistics challenges.* Available at https://www.atkearney.com/documents/10192/253176/Chinas_E-Commerce_Market.pdf

Abad, A. (2015, January 10). *India's economy, The Gujarat model.* The Economist. Accessed February 01, 2016, from http://www.economist.com/news/finance-and-economics/21638147-how-modi-nomics-was-forged-one-indias-most-business-friendly-states

African Development Bank Group. (2016). *Nigeria economic outlook.* Accessed February 04 2016, from http://www.afdb.org/en/countries/west-africa/nigeria/nigeria-economic-outlook/

African Economic Outlook, Egypt. (2015). Accessed February 03, 2016, from http://www.africaneconomicoutlook.org/en/country-notes/north-africa/egypt/

Alsahary, I. (2015, December 20). *Egypt's non-petroleum exports drop 19 percent in November 2015.* Egypt Independent. Accessed February 03, 2016, from http://www.egyptindependent.com/news/egypt-s-non-petroleum-exports-drop-19-november-2015

ASEAN. (2015a). ASEAN Economic community at a glance. Available at http://www.asean.org/storage/2015/12/AEC-at-a-Glance-2015.pdf

ASEAN. (2015b). *AEC blueprint 2025.* Available at http://www.asean.org/storage/images/2015/November/aec-page/AEC-Blueprint-2025-FINAL.pdf

Asian Scientist Magazine. (2013, May 6). *China to build billion-dollar robot industrial base in liaoning.* Accessed February 05, 2015, from http://www.asianscientist.com/2013/05/topnews/china-build-us8-billion-robot-industrial-base-2013/

Australian Government; Department of Foreign Affair and Trade. (2016). *MIKTA—Mexico, Indonesia, the Republic of Korea, Turkey, Australia.* Accessed April 04, 2016, from http://dfat.gov.au/international-relations/international-organisations/mikta/Pages/mikta.aspx

Avasthy, A., Dekhne, A., Malik, Y., & Weydringer, J. (2015, January). Winning supply chain strategies for African markets. *Supply Chain Quarterly.* http://www.supplychainquarterly.com/topics/Strategy/20150331-winning-supply-chain-strategies-for-african-markets/ Accessed 9/26/2016

Bajpai, P. (2014, October 31). *China's GDP examined: A service-sector surge.* Investopedia. Accessed February 02, 2016, from http://www.investopedia.com/articles/investing/103114/chinas-gdp-examined-servicesector-surge.asp

Bhargava, Y. (2015, February 23). *India has second fastest growing services sector.* The Hindu. Accessed February 01, 2016, from http://m.thehindu.com/business/budget/india-has-second-fastest-growing-services-sector/article6193500.ece

Bilby, E. (2014, April 1). *Losing mileage.* Breakingviews. Accessed February 02, 2016, from http://www.breakingviews.com/chinas-car-joint-ventures-arent-built-to-last/21140977.article

Biller, D. (2016, January 25). *Brazil's highs and lows.* Bloomberg. Accessed February 01, 2016, from http://www.bloombergview.com/quicktake/brazils-highs-lows

Blodget, H. (2012, January 22). *This article explains why Apple makes iPhones in China and why the US is screwed.* Business Insider. Accessed February 05, 2015, from http://www.businessinsider.com/you-simply-must-read-this-article-that-explains-why-apple-makes-iphones-in-china-and-why-the-us-is-screwed-2012-1

Bloomberg. (2016, January 25). *Brazil's highs and lows.* Bloomberg Quick Take. Last accessed February 01, 2016, from http://www.bloombergview.com/quicktake/brazils-highs-lows

BMW. (2016). *Manufacturing facilities, Rosslyn plant.* Accessed February 02, 2016, from http://www.bmwgroup.com/e/0_0_www_bmwgroup_com/produktion/produktionsnetzwerk/produktionsstandorte/werk_rosslyn.html

Canuto, O. (2016, January 26). *Has the global trade-development link peaked?* The Huffington Post. Accessed September 15, 2016, from http://www.huffingtonpost.com/otaviano-canuto/has-the-global-trade-deve_b_9077698.html

Chinadaily.com. (2015, December 24). *769b yuan invested in developing western China in 2015.* Accessed February 02, 2016, from http://usa.chinadaily.com.cn/business/2015-12/24/content_22793247.htm

Colombo, J. (2014, May 19). *A guide to South Africa's economic bubble and coming crisis.* Forbes. Accessed February 02, 2016, from http://www.forbes.com/sites/jessecolombo/2014/03/19/a-guide-to-south-africas-economic-bubble-and-coming-crisis

Deccan, H. (2016, February 8). *Global maritime meet soon, says Modi.* Accessed February 10, 2016, from http://www.deccanherald.com/content/527634/global-maritime-meet-soon-says.html

Demand Caster. (2010, May 4). *The Eyjafjallajökull effect.* Accessed January 15, 2016, from https://demandcaster.com/2010/04/2010430the-eyjafjallajokull-effect-html/

Democracy Corps. (2016, January 26). *Taking on trade, unions and banks in 2016.* Accessed September 15, 2016, from http://www.democracycorps.com/Rewriting-the-Rules-of-the-Economy/taking-on-trade-unions-and-banks-in-2016/

Deshpande, R. (2015, December 26). *India likely to top world's growth chart: Harvard study.* The Times of India. Accessed February 01, 2016, from http://timesofindia.indiatimes.com/india/india-likely-to-drop-worlds-growth-chard-harvard-study/articleshow/50328665.cms?from=mdr

Erb Phillips, K. (2016, September 5). *So you want to open a taco truck? 11 steps to get started with your small business.* Forbes. Accessed September 15, 2016

Erickson, J. (2011, January 20). *Alibaba Group calls for logistics revolution.* Alizila.com. Accessed February 02, 2016, from http://www2.alizila.com/alibaba-group-calls-logistics-revolution

Ernst & Young. (2014). *The road to 2030: A survey of infrastructure development in Russia* (p. 8). London.

Exporters' Assembly of Turkey (TİM). (2016). http://www.tim.org.tr/en/

Financial Post. (2013, February 1). *Daimler takes stake in BAIC Motor with eye on China's growing luxury car market.* Last accessed February 01, 2016, from http://business.financialpost.com/news/transportation/daimler-takes-stake-in-baic-motor-with-eye-on-chinas-growing-luxury-car-market

Florida, R., Mellander, C., & King, K. (2015). *The Global Creativity Index 2015.* Martin Prosperity Institute, University of Toronto

FM Global. (2015). *Resilience Index 2015.* Available at http://www.fmglobal.com/assets/pdf/Resilience_Methodology.pdf

Halley, A., & Torné, R. (2015, July 23). *ASEAN: Economic progress and integration.* CEIC Data. Accessed September 15, 2016, from http://www.ceicdata.com/en/press/asean-economic-progress-and-integration

Homeland Security News Wire. (2012 January). *Disaster insurance: 2011 natural disasters cost U.S. insurers more than $32 billion.* Accessed January 15, 2016, from http://www.homelandsecuritynewswire.com/dr20120103-2011-natural-disasters-cost-u-s-insurers-more-than-32-billion

Hurriyet Daily News. (2013, September 5). *Transatlantic alliance to cost Turkey '$20 billion'.* Accessed April 04, 2016, from http://www.hurriyetdailynews.com/transatlantic-alliance-to-cost-turkey-20-billion.aspx?pageID=238&nid=53844

Hurriyet Daily News. (2016, February 01). *Turkey's exports drop 14.4 pct in first month of 2016.* Accessed April 04, 2016, from http://www.hurriyetdailynews.com/turkeys-exports-drop-144-pct-in-first-month-of-2016.aspx?pageID=238&nID=94596&NewsCatID=344

Indian Brand Equity Foundation (IBEF). (2015). *Ports in India.* Accessed February 01, 2016, from http://www.ibef.org/industry/ports-india-shipping.aspx

International Labour Organization. (2014, August 19). *Can the AEC 2015 deliver for ASEAN's people?* Accessed September 15, 2016, from http://www.ilo.org/asia/info/public/features/WCMS_300671/lang--en/index.htm

International Monetary Fund, Western Hemisphere Dep. (2015). *Colombia: Selected issues paper* (p. 39)

Kauffmann Bossart, M. (2015, April 7). *Türkisch-syrisches Grenzgebiet: Wirtschaft im Sog der Kriegswirren.* Neue Zürcher Zeitung. Accessed April 04, 2016, from http://www.nzz.ch/wirtschaft/syrische-tageloehner-und-tuerkische-kriegsgewinnler-1.18517532

Kortam, H. (2015, November 22). *Egypt's tourism in crisis.* Egypt Independent. Accessed February 03, 2016, from http://egyptianstreets.com/2015/11/22/egypts-tourism-in-crisis-number-of-tourists-to-drop-by-13-percent/

KPMG. (2011, May). *On the move in China. The role of transport and logistics in a changing economy* (p. 14 ff)

KPMG International. (2010). *Global transport logistics in India, Part 2.* Accessed February 01, 2016, from http://www.slideshare.net/srinivasangsrinivasan/logistics-inindiapart2

KPMG, KPMG South Africa Blog. (2013). *Challenges facing supply chains in South Africa.* Accessed February 03, 2016, from http://www.sablog.kpmg.co.za/2013/12/challenges-facing-supply-chains-south-africa/

Laursen, W. (2015, May 5). *Vietnamese container ports top 2014 growth.* The Maritime Executive. Accessed September 15, 2016, from http://www.maritime-executive.com/article/vietnamese-container-ports-top-2014-growth

Lidegaard, B. (2016, July 25). *Hillary Clinton and the Scandinavian-American dream.* Project Syndicate. Accessed September 15, 2016, from https://www.project-syndicate.org/commentary/scandinavian-welfare-economies-success-by-bo-lidegaard-2016-07

Mello, J. (2012, November 12). *Logistics and distances in Brazil.* The Brazilian Business. Accessed February 01, 2016, from http://thebrazilbusiness.com/article/logistics-and-distances-in-brazil

Meyer, P. J. (2014, March 27). *Brazil: Political and economic situation and U.S. relations.* Congressional Research Service

Mullen, J. (2016, June 23). *China can't buy enough industrial robots.* CNN. Accessed September 15, 2016, from http://money.cnn.com/2016/06/23/technology/china-industrial-robots/

OECD. (2015). *OECD Economic Survey Indonesia 2015* (pp. 73 and 64)

PricewaterhouseCoopers. (2015, January). *A decade of unprecedented growth—China's impact on the semiconductor industry.* Available at https://www.pwc.com/il/en/assets/pdf-files/china-semicon-2014.pdf

PwC Economics & Policy/Nigeria Economy Watch. (2015, May). *What next for Nigeria's economy? Navigating the rocky road ahead.* Available at https://www.pwc.com/ng/en/assets/pdf/economy-watch-may-2015.pdf

Roland Berger. (2014, August). *Global logistics markets* (p. 20)

rt.com. (2016, January 21). *Dubai to invest $2bn in Russian ports.* Accessed March 31, 2016, from https://www.rt.com/business/329687-dubai-russia-ports-joint/

SAIC USA Inc./PR Newswire. (2012, June 27). *Chinese auto giant SAIC Motor celebrates opening of new North American Operations Center in Michigan.* Accessed Febryary 02, 2016, from http://www.prnewswire.com/news-releases/chinese-auto-giant-saic-motor-celebrates-opening-of-new-north-american-operations-center-in-michigan-160541035.html

Schwab, K. (2015). *The global competitiveness report 2015–2016.* World Economic Forum. Available at http://www3.weforum.org/docs/gcr/2015-2016/Global_Competitiveness_Report_2015-2016.pdf

Shih, G. (2015, September 20). *Technology gap gives foreign firms the edge in China robot wars.* Reuters. Accessed February 09, 2015, from http://www.reuters.com/article/china-robots-idUSL1N11N0C220150920

Sim, E. (2015, October 7). *The TPP's impact on ASEAN.* ASEAN Economic Community Blog. Accessed Sepetmber 15, 2016, from http://aseanec.blogspot.co.id/2015/10/the-tpps-impact-on-asean.html

Simoes, A. (2016). *Coal Briquettes.* The Observatory of Economic Complexity (OEC). Accessed September 15, 2016, from http://atlas.media.mit.edu/en/profile/hs92/2701/

Southafrica.info. (2015, December 4). *South Africa, China sign trade and industry deals.* Accessed February 03, 2016, from http://www.southafrica.info/business/Chinese-South-Africa-trade-041215.htm

Statista—The statistical portal. (2016). *Gross domestic product (GDP) of the BRIC countries from 2010 to 2020 (in billion U.S. dollars).* Accessed February 01, 2016, from http://www.statista.com/statistics/254281/gdp-of-the-bric-countries/

Taylor, E., & Shen, S. (2014, August 25). *China tells German auto suppliers to find partners.* Reuters. Accessed February 01, 2016, from http://uk.reuters.com/article/uk-autos-china-partnership-idUKKBN0GP0M720140825

The Economist. (2013, May 30). *Towards the end of poverty.* Accessed September 15, 2016, from http://www.economist.com/news/leaders/21578665-nearly-1-billion-people-have-been-taken-out-extreme-poverty-20-years-world-should-aim

The Observatory of Economic Complexity (OEC), India. (2016). Accessed February 01, 2016, from http://atlas.media.mit.edu/en/profile/country/ind/

The Republic of Turkey Prime Ministry, Investment Support and Promotion Agency, Foreign Trade. (2016). Accessed April 04, 2016, from http://www.invest.gov.tr/en-US/investmentguide/investorsguide/Pages/InternationalTrade.aspx

The World's richest countries. (2016). Accessed February 09, 2016, from http://www.worldsrichestcountries.com/top_indonesia_exports.html

Thompson, C. (2015, March 25). *Drive from Europe to the U.S.? Russia proposes world's greatest superhighway.* CNN. Accessed March 31, 2016, from http://www.cnn.com/2015/03/24/travel/trans-siberian-road/

Trading Economics. (2016a). *Brazil's exports, 1954-2016.* Accessed February 01, 2016, from http://www.tradingeconomics.com/brazil/exports

Trading Economics. (2016b). Colombia. Accessed February 03, 2016, from http://www.tradingeconomics.com/colombia/gdp-growth

United Nations Conference on Trade and Development (UNCTAD). (2016, April 14). *New database of ASEAN non-tariff measures.* Accessed September 15, 2016, from http://unctad.org/en/pages/newsdetails.aspx?OriginalVersionID=1234&Sitemap_x0020_Taxonomy=UNCTAD%20Home;#2030

Villarreal, M. A. (2014, February 14). *The U.S.-Colombia free trade agreement: Background and issues.* Congressional Research Service. Available at https://www.fas.org/sgp/crs/row/RL34470.pdf

Vollgraaff, R. (2015, October 1). *U.S.-South Africa trade spat risks $1.7 billion of exports.* Bloomberg Business. Accessed February 02, 2016, from http://www.bloomberg.com/news/articles/2015-10-01/u-s-south-africa-trade-dispute-risks-1-7-billion-of-exports

Wikipedia. (2016a). *Automotive industry in China.* Accessed February 02, 2016, from https://en.wikipedia.org/wiki/Automotive_industry_in_China

Wikipedia. (2016b). *Pacific alliance*. Accessed February 03, 2016, from https://en.wikipedia.org/wiki/Pacific_Alliance

World Bank. (2014). *Connecting to compete 2014, trade logistics in the global economy—The logistics performance index and its indicators*. Available at http://www.worldbank.org/content/dam/Worldbank/document/Trade/LPI2014.pdf

World Bank. (2015, October 4). *World Bank forecasts global poverty to fall below 10% for first time; Major hurdles remain in goal to end poverty by 2030*. Accessed September 15, 2015, from http://www.worldbank.org/en/news/press-release/2015/10/04/world-bank-forecasts-global-poverty-to-fall-below-10-for-first-time-major-hurdles-remain-in-goal-to-end-poverty-by-2030

World Bank (2016a). *Data, India*. Accessed February 01, 2016, from http://data.worldbank.org/country/india

World Bank. (2016b). *Data, Turkey, annual GDP growth (%)*. Accessed April 04, 2016, from http://data.worldbank.org/country/turkey

World Bank. (2016c). *GDP growth (annually %)—Russian Federation; 1996–2014*. Accessed March 23, 2016, from http://data.worldbank.org/indicator/NY.GDP.MKTP.KD.ZG

World Bank. (2016d). *Global rankings 2016*. Accessed September 15, 2016, from https://lpi.worldbank.org/international/global

World Bank. (2016e). *Indonesia*. Accessed February 03, 2016, from http://www.worldbank.org/en/country/indonesia/overview

World Bank. (2016f). *Vietnam*. Accessed February 03, 2016, from http://www.worldbank.org/en/country/vietnam/overview

World Bank. (2016g). *World development indicators, Brazil*. Accessed February 01, 2016, from http://data.worldbank.org/country/brazil

World Trade Organization. (2012). *Statistics: International trade statistics 2012, Chart 1 Volume of world merchandise exports and gross domestic product, 1950-2011*. Available at https://www.wto.org/english/res_e/statis_e/its2012_e/charts_e/chart01.pdf

Yang, J. (2011, October 20). *Worst Thai flood in 50 years hit Apple, Toyota supply chain*. Bloomberg. Accessed January 05, 2016, from http://www.bloomberg.com/news/articles/2011-10-20/worst-thai-floods-in-50-years-hit-apple-toyota-supply-chains

Yinan, G. (2016, February 29). *Express delivery changing China's consumption map*. People's Daily. Accessed March 04, 2016, from http://en.people.cn/n3/2016/0229/c98649-9022848.html

Zuvin, S., & Kurtuldu, A. (2015, February 20). *Economic impact of TTIP on the US, EU and Turkey*. Mondaq. Accessed April 04, 2016, from http://www.mondaq.com/turkey/x/376266/international+trade+investment/Economical+Impacts+Of+TTIP+On+The+US+EU+And+Turkey

Global Dynamics and Key Trends

3

For supply chain managers, long-term considerations regarding economic development, the future of industrial sectors and ecosystems, and geopolitics were less of a priority in the past. Events, such as 9/11, the Arab Spring, and the growing number of trade barriers and geopolitical tensions impacting the global supply and value chain, however, have stressed the importance of monitoring key trends within the economic environment.

3.1 Megatrends: Forces Impacting Everyone

The so-called megatrends are greatly influencing ways of thinking and acting. The term refers to long-term developments that affect all areas of society and economy. Many of these trends impact the performance of supply chains. Policymakers, businesses, and citizens might not even realize they have come under the influence of a megatrend. Therefore, CSCOs have to take the impact of various megatrends into consideration when planning and managing the supply chain ecosystem.

3.1.1 Business Ecosystems

Effective collaboration is essential in order to thrive and succeed in the new ecosystem world. James F. Moore originated the strategic planning concept of a business ecosystem, now widely adopted in the tech community. He defines the business ecosystem as follows:

> *An economic community supported by a foundation of interacting organizations and individuals—the organisms of the business world. The economic community produces goods and services of value to customers, who are themselves members of the ecosystem. The member organisms also include suppliers, lead producers, competitors, and other stakeholders. Over time, they coevolve their capabilities and roles, and tend to align themselves with the directions set by one or more central companies. Those companies*

© Springer International Publishing AG 2017
W. Lehmacher, *The Global Supply Chain*, Management for Professionals,
DOI 10.1007/978-3-319-51115-3_3

holding leadership roles may change over time, but the function of the ecosystem leader is
valued by the community because it enables members to move toward shared visions to
align their investments, and to find mutually supportive roles. (Moore 1996)

Ecosystems are dynamic and coevolving communities of diverse players who create and capture new value through both collaboration and competition (Deloitte University Press 2015). The invention and expansion of the Internet had a major impact on traditional boundaries. Digitalization allows businesses to form integrated mutually beneficial partnerships, some long term, some over a short period of time. Business ecosystems profit from covering more steps of the supply and value chain than most single companies would be able to establish alone. Due to complementary competences and structures and the high level of autonomy and decision-making authority of the various participants, business ecosystems are able to adapt to changes and shocks faster than large enterprises.

The supply and value chain can be understood as a well-integrated business ecosystem. The Internet and digital platforms also allow entrepreneurs and companies in rural areas to start and drive global businesses, for example, by producing through contract manufacturers in Asia and selling via one of the large e-commerce platforms. Small business owners can grow large by turning to the global community via the Internet. For example, Ryan French, a student of Washington University designed the GameKlip™, a device that attaches your Android phone to a Dual Shock 3 controller, normally used for the Play Station 3. This allows gamers to use a real controller to play games on a smartphone. French sold the controller over the Internet in 80 countries; most GameKlips were not even shipped to US residents, but to gamers in Canada, Great Britain, Australia, and Germany (GameKlip 2016).

Within the ecosystem, new forms of cooperation evolve. One of those is "coopetition": competitors cooperate to reach higher-value creation.[1] Instead of being limited by their own core competence, businesses try to stretch their scope by leveraging competences of partners and competitors. New organizational structures emerge. Companies can even go beyond the frontiers of their traditional sector into new business fields, creating horizontal partnerships and driving business convergence.

For example, in 2014, *DHL* acquired *StreetScooter*, a spin-off by Institute of Technology Aachen and a consortium of approximately 80 industrial companies in the automotive industry and related sectors, aimed at developing affordable electric vehicles. Will this acquisition turn the postal and logistics group into a car manufacturer? Probably not, but it ensures access to a type of vehicle DHL requires to conduct its business more effectively and build competitive edge.

Is *Google* a technology company or a mobility group? With its engagement in the self-driving vehicle business, one could raise this question. The search engine giant is on its way to entering not only the passenger transportation market, but

---

[1]Term used to describe cooperative competition.

potentially also the logistics business. The company requested a patent for an autonomous delivery vehicle, which was awarded in February 2016. The patent describes a vehicle that would be used to transport packages and parcels from a pickup point to delivery destinations in secured compartments (Wolfgang 2016).

Amazon also seems to be moving into global shipping and logistics: expanding from its "Fulfillment by Amazon" service, which provides storage, packing, and shipping for independent merchants selling products on the company's website, to a global delivery operation (Soper 2016). The project "Supply Chain by Amazon" aims to control the flow of goods from factories in China and India to customers' doorsteps in New York, London, and Paris. This entry could cause a threat to established delivery companies and might restructure the entire delivery market over time. Amazon has already offered domestic services that go beyond their original core business and competence such as the installation of appliances. Called Home Service, the offering is adding a new dimension to the existing range of services. In 2016, the conglomerate announced having signed a deal to lease 20 Boeing 767 wide-body freighter aircraft to handle more of its own deliveries in the USA. Amazon has the volume to take supply chain management in own hands. According to Reuters, Amazon spent $11.5 billion on shipping in 2015 (Sachin R and Saito 2016). Once in operation, the logistics system could be the platform to offer logistics services to third parties, in particular to small and midsized businesses.

How Business Ecosystems Impact Global Supply Chains
Due to digitalization, the world of business is increasingly interconnected. That applies to the companies within the ecosystems as well as the ecosystems themselves, which link to more and more other systems. With the help of these links, companies can more easily build and leverage knowledge to enter adjacent fields. As a result, boundaries between industries and even systems have started to blur. Foresight, imagination, vision, and new strategies are required to navigate in the world of business ecosystems. The global supply chain is an ecosystem itself. The global supply and value chain ecosystem includes global sourcing and procurement, decentralized production or production networks, and global distribution and repurposing operations, such as the activity in secondary markets.

Participants within specific ecosystems are encroaching on each other's business areas. For example, supply chain service providers, in addition to handling, warehousing, transportation, and return logistics, may offer quality control, assembly of parts, and recycling. By mastering the management of assets and the creation of value in areas that are outside the traditional boundaries, supply chain managers create what Deloitte University Press calls value webs (Marchese and Kelly 2015). In the future, the ability to understand one's own and adjacent systems, collaborate with partners outside the ecosystem, and expand into new fields will be crucial for success, differentiation, and survival. As much as companies will need to explore new fields, they will also have to deal with new players entering their turf. Google entering the mobility market is just one example.

Shareholders, management, employees, suppliers, and customers are direct stakeholders of the ecosystem; unions, citizens, and the media are indirect stakeholders. All parties need to be aligned by a shared vision and find a common modus operandi. Comprehensive stakeholder management is therefore another critical competence in the era of business ecosystems.

Some ports have realized the importance of the supply chain ecosystem as a differentiating factor and facilitate infrastructure investments. The Netherlands invested heavily in its infrastructure around the port of Rotterdam and was rewarded with being named "best port infrastructure in the world," according to the World Economic Forum's Global Competitiveness Report 2015–2016. Singapore came second, and the United Arab Emirates held onto third place.

The EU announced in 2016 that they are intending to invest just shy of 13 billion EUR ($14.2 billion) in infrastructure projects by 2020. The financing will pay for expanding networks in the transport, communication, and energy sector (Spirkl 2016). In emerging markets, according to the Financial Times, building infrastructure fit for the twenty-first century is forecast to require a trillion dollars a year in additional financing (Flood 2015).

3.1.2 Globalization X.0

Globalization will continue. Capital and goods, data and information, and talent and culture will continue to zigzag across the globe. Strategy consultants at Roland Berger predict that by 2030, worldwide exports will have more than tripled; exports from emerging and developing economies will have more than quadrupled. They also expect further boosts in worldwide trade due to regional and bilateral agreements (Roland Berger 2014). Global encounter and exchange is driven by trade agreements, favorable shipping costs, and growing access to the Internet. Worldwide collaboration and the number of value webs are constantly increasing, not only on the business side. However, there has been a shift in economic power. Emerging economies have become more participative and influential in the global economy.

The BRIC states are often referred to as "the new powerhouse." The share of worldwide GDP of the BRIC countries will increase from 27% in 2013 to 38% in 2030. As indicated above, however, this growth is not homogeneous: China and India are likely to grow stronger within the world GDP, while Brazil and Russia have already fallen behind. Other emerging markets, such as MINT, CIVETS, and the so-called Next Eleven (Bangladesh, Egypt, Indonesia, Iran, Mexico, Nigeria, Pakistan, Philippines, Turkey, South Korea, and Vietnam), are also expected to play an increasingly important role in the global economy. While the developing world is on the rise, developed economies find more competition. A certain leveling is taking place, making the world a more inclusive place.

Not only nations but also large cities will play a decisive role in future globalization. Strong connections between leading cities will be the pillars for the platforms of new economic blocs. Local governments are more flexible and often equipped with more decision-making power than their federal counterparts. As the economic center of trade and world GDP shifts eastward, so will the growth of urban areas. Delhi, Shanghai, Mumbai, and Beijing will be among the top urban agglomerations. In 1950, 22% of the world's top 50 urban agglomerations were located in the USA. By 2030, only 6% of the world's urban commercial centers are predicted to be there (Credit Suisse Research Institute 2015).

New logistics hubs have already emerged. The magazine *Supply Chain* rated three airports in emerging countries among the top five airfreight hubs in the world. It voted Hong Kong International Airport first, with a freight volume of nearly 3.7 million tons. The next runner up was the FedEx Super Hub at Memphis International Airport, which has the largest cargo operation by volume of any airport worldwide – nearly 3.7 million tons annually. Shanghai Pudong International Airport cargo terminal in China (with a freight volume of nearly 2.6 million tons per year); Incheon International Airport, South Korea (2.7 million tons a year); and Frankfurt Airport, Germany (over 2.1 million tons of freight a year), take up the subsequent places (Ryan 2012).

Due to a higher demand for regional and global supply chain platforms, logistics companies are establishing international networks. One example is the *IFLN network*, established in 2000 and headquartered in Houston, TX. It consists of 260 specialist freight companies in more than 100 countries and aims to "provide comprehensive supply chain solutions and end-to-end logistics from project initiation to sustainable operations" (IFLN Network 2016). With the exception of large markets, such as the USA, China, and India, the network's goal is to offer a choice of at least three members per country or main city providing a range of services and capabilities.

Another example for pooling intercontinental capacities and capabilities is the alliance of shipping companies *Maersk Line*, based in Copenhagen, Denmark, and *Mediterranean Shipping (MSC)*, an unlisted Swiss-based company. Among them, they pool 185 ships on European, transatlantic and transpacific routes, saving Maersk Line $350 million a year, according to its own calculations. The alliance covers 35% of the Asia-Europe route, 31% of transatlantic trade, and 22% of transpacific (Wallis and Zawadzki 2014). Following the collapse of Hanjin Shipping (The Economist 2016), a South Korean container line, on August 31, 2016, investors are now pressing for other shipping lines to merge as even the strongest ocean carriers are suffering.

The rebalancing has taken place. The world might have moved beyond Globalization 2.0 towards a web of connected cities in a new era of global trade. Main characteristics of the new era of global trade are relatively mature value and supply chains, new protectionism, and the rise of nontariff barriers after a period of significant tariff reductions, increasing regionalization and localization of production, and the boom of e-commerce, which struggles to overcome the hurdles in international business (Murray 2016).

Digitization appears as the key enabler of modern times and an important lever to capture value in the new era of global trade. It is digitization which opens up new horizons for product design and management, manufacturing, retail, after-sales, and the repurposing of goods. We need open and courageous leaders in the public and private sector to make the change happen and capture the opportunity in the following main areas of development.

The New Customer Experience: Fast Products
Today's customers demand fast products, such as fast fashion, with quickly changing models. In fact, this goes well with the interests of the brands, avoiding high inventory and high risks resulting from making bets on the right models for an entire fashion season, for instance. Fast products require short supply chains. Consequently, production needs to move closer to the markets and shops in order to meet the shorter lead times from sketch to shelf. Zara, for example, "adapts couture designs, manufactures, distributes, and retails clothes within 2 weeks of the original design first appearing on catwalks" (Lu 2014). In creating the new customer experience, the management of the supply chain becomes a critical source of competitive edge.

Companies that wish to play in the "fast economy" will require new factories close to the markets and new distribution platforms. The countries which provide, among other things, the most fluid import and export ecosystem will be high on the list for investments. In particular, countries in regions close to the large markets, such as Central America, Southeast Asia, and Northern Africa, might wish to review their strategies to capture the value of this trend. With the fast economy, some intercontinental flows of goods will become regional and local traffic. However, the new factories will continue to require global supply as not all materials and resources will come from the sources nearby.

Multilayer Global Manufacturing and Supply Platforms
Digitization helps the fast economy. In the past, brands tended to centralize production for better manageability and quality control. The power of information technology, the Internet of Things, big data, and the cloud provide a new level of collaboration and empowerment throughout the value and supply chain. One example is the Flex Pulse Center (Flextronics 2015). The enhanced visibility allows companies to move the factories closer to the customer without risking sudden surprises. In the continuous process, the global supply chain is converting to a more and more dense and integrated platform of short-, medium-, and long-distance cargo transfers with regional and local distribution centers along the way.

These multilayer global manufacturing and supply platforms still have many black holes. Some blackouts of visibility are caused by governmental security concerns, for example, in free zones, and others by the lack of digital infrastructure. There are opportunities for business and government, through public-private partnerships, to establish not only the digital architecture but also the trust to fully leverage the available technology. It goes without saying that the smooth and seamless movements of goods in and out of countries through the reduction or elimination of tariff and nontariff border barriers are a prerequisite for establishing denser and more cost-optimized platforms and lifting the attractiveness of countries and locations.

Scaling Up Market Participation: International E-Commerce

Platformization allows for a more inclusive economy. Global e-commerce platforms can connect millions of manufacturers and billions of consumers, offering the global market to even the smallest manufacturer and the broadest choice to any consumer (Ja-young 2015). eBay and Alibaba are examples of this. In addition to much broader and better matchmaking, middlemen are cut out, which allows for higher margins on the sales and lower prices on the purchasing side. Furthermore, risks of corruption are reduced too. However, the concept only works provided the underlying logistics and transportation platforms support the digital transactions.

Unlocking the potential of international e-commerce requires efficient and cost-effective logistics and smooth customs processes through the paperless digital export and import processing. On the import side, there is a need for effective tools to be able to process and analyze information about shippers and products moved. These tools reduce clearing times, even enable preclearance, and help to manage the risks which come along with the lower-value goods which often fall below the threshold of more diligent customs clearance processes. In addition, governments need to ensure healthy competition and avoid the formation of digital monopolies. Platforms can also be used to foster the job-creating small and mid-sized business landscape.

Repurposing of Goods

Although there is a need to analyze the entire value chain, since sometimes even long-distance transport might be less carbon intense than local production, tightening the supply chain in many cases saves energy and emissions. Hence, the trend towards localization and regionalization helps with resources and the climate. However, new technology can go much beyond the simple shortening of the supply chain.

The new visibility in the supply chain not only helps to identify leaks and misuse, but enables new operating and business models, ranging from optimizing delivery routes to tapping into unused capacity enabled by the many platforms of the sharing economy. Additional potential lies in the resources locked in the products which are thrown away every day: electronics, paper, and plastics. Asset tracking could help unlock a potential value of about $52 billion annually for consumer electronics and household appliances alone (Ellen Macarthur Foundation 2015). Repurposing products will be a major challenge and opportunity for governments and companies in the years ahead.

In the modern interconnected and interdependent world, we need seamless global visibility and fluidity of the flow of goods. Businesses seek and will find new sources of value in tracking products throughout the entire first life cycle and recovery for repurposing. Fast products represent particular challenges. Through creating a repurposing capability, the public and private sectors have a unique opportunity to position themselves as modern and responsible players.

Developed and developing countries have benefitted from globalization, with China as the most recent role model of trade-driven growth. Some low-cost labor countries might still be able to repeat China's success, while others need to look for new models. While globalization has driven the rise of the emerging markets and global players, it has also paved the way for today's dense multilayered value and supply platforms which are the basis of our modern life. Digitization is enabling new visibility to bring globalization to the next level.

How Globalization X.0 Impacts Global Supply Chains

In recent years, the global economy has experienced the emergence and growths of value networks. Repartition of work across the globe has been a strong driver of trade. This led to today's globally integrated multilayered supply and value chain.

As emerging markets continue to grow, CSCOs will no longer be able to rely only on national logistics service providers. In the future, they will have to outsource to logistics specialists that are able to provide global platforms for multi-country standardized supply, distribution, and after-sales services including repurposing. Thus, most of the leading logistics and supply chain service providers invest in the expansion of their networks, create new hubs and trade lanes, and strengthen global, regional, and domestic platforms. Air and sea freight in particular are moving towards more concentrated networks, with large platforms and hubs at the regional and even national level.

Digitization, platformization, the fast economy, and new protectionism create new opportunities and challenges for the supply and value chain. Players need to establish nimble networks to be prepared and ready to act in any situation. Monitoring of the ecosystem evolution and its environment is a prerequisite for survival.

Multinational companies are increasingly building factories in local markets; General Electric, for example, has started to make engine parts where they are needed rather than shipping them from the USA to the world (The Economist 2016). This is not an exception, but part of a radically new company strategy. In May 2016, CEO Jeffrey Immelt said that General Electric "is making a bold pivot in strategy as a response to rising protectionist political pressures" (Murray 2016). Pressure applied by various stakeholders induces global logistics service providers to offer more services in a broader scope to be able to provide national, regional, and global coverage. Future supply chain service providers and their CSCOs will be both global citizens and local champions.[2]

3.1.3 Population Growth and Demographics

The United Nations projects that world population will reach 8.5 billion by 2030 and increase further to 9.7 billion by 2050 and 11.2 billion by 2100. More than half of global population growth between 2015 and 2050 is expected to occur in Africa;

---

[2]The term global or world citizenship or world typically refers to a person who places their identity with a "global community" above their identity as a citizen of a particular nation or place. A local champion is an individual that is viewed as popular, able, and excellent within a relatively small area.

4.2 billion people will live in Africa by 2100, quadrupling the population of 2015. Behind Africa, Asia is projected to be the second largest contributor to future global population growth, followed by North America, Latin America, and the Caribbean and Oceania, which are projected to have much smaller increments. The U.S population is expected to grow from 310 million to 439 million between 2010 and 2050, an increase of 42%. Forecasts for the Canadian population also see continuous growth of approximately 1% annually. In 2056, the U.S neighbor could be home to 42.5 million people (from 32.6 million in 2006). In contrast, Europe is likely to have a smaller population in the medium variant in 2050 than in 2015 (United Nations 2015).

Significant gain in life expectancy is also projected. Globally, life expectancy at birth has reached 70 years and is likely to rise to 77 years by 2045–2050 and to 83 years by 2095–2100. The population in today's emerging countries will be younger than the average age of people in industrial countries. In 2050, the number of Americans aged 65 and older is projected to be 88.5 million, more than double its population of 40.2 million in 2010. Some of the older people might still be working, while others either live a life of self-determination that requires mobility or will have to rely on long-term care. Each of which demands a specific supply chain.

New business strategies and services will emerge with the increase in life expectancy. Elderly people require specific products, such as medications, foods, healthcare products, and a range of different services. This could lead to a rising number of touchpoint services, and enriched interfaces between consumer and company. Delivery personnel might provide care services, and phone hotlines help in emergencies. Technology will play a key role, and the IoT will bring comfort and safety.

Another population trend to note is the growing number of immigrants to North America and Europe. The USA will become more racially and ethnically diverse, with the aggregate minority population projected to become the majority in 2042 (United States Census 2010). In Canada and Germany, migration is the source of required additional workforce. In Germany, about 300,000 net immigrants per year would be required for the next 40 years (compared with just 13,000 in 2009) in order to maintain its current GDP growth rates through 2050 (Boston Consulting Group 2011). In Canada, international migration accounted for two-thirds of the country's population growth in 2006; around 2030, deaths are expected to start outnumbering births. From that point forward, immigration would be the only growth factor for the Canadian population (Statistics Canada 2016). Between 2015 and 2050, the top net receivers of international migrants (more than 100,000 annually) are projected to be the USA, Canada, the UK, Australia, Germany, the Russian Federation, and Italy, according to the United Nations.

How Population Growth and Demographics Impact Global Supply Chains
The world's population is growing fast, and life expectancy is rising. More people require more supply, such as expanding demands for healthcare and life science logistics. UPS estimates that healthcare logistics is a $70 billion global market, growing at an annual rate of 4% (Brennan and Golden 2015). Transport

infrastructure is saturated. While travel and transportation are on the rise, drivers and pilots are in short supply in many countries. Boeing has estimated a global shortage of 533,000 pilots over the next 20 years (Diehl 2015).

Automation is one way to deal with the shortage of labor. Automated guided vehicles (AGVs) are already in place in many warehouses; warehouses run by retailers *Gap*, *Zappos*, and *Staples* use autonomous robots to pick products from the shelves and send them out. According to the "Trend Report: Robotics in Logistics" by DHL, some 15% of current warehouses have already been mechanized. Robot technology will make the supply chain "faster, safer, and more productive," the authors claim. Robot or robotic technology will improve all areas of logistics.

In warehouses and distribution centers, we will see goods-to-picker systems, mobile piece picking, exoskeletons, swarm robots, and automated deliveries by drones and eventually autonomous vehicles; the hubs will run 24/7 to better align with distribution centers and local delivery operations. Supply chain service providers will employ robots to carry parcels to the loading bay, load and unload trucks and containers, or handle dangerous goods. Last mile delivery will see service robots that deliver parcels or help delivery personnel to carry heavy loads (Bonkenburg 2016). We will see delivery vehicles combined with onboard drones and robots.

Amazon was the first company to use robots in warehouses at large scale. In September 2015, the company announced it would have 30,000 mobile robotic fulfillment systems, Kiva robots, at work in 13 fulfillment centers by the end of the month. This effectively doubled the number of bots they had installed so far (Tobe 2015). In Stockholm Arlandastad, the logistics group *DB Schenker* has implemented the automated picking and return system CarryPick Swisslog for the Swedish toy merchant *Lekmer*. Powerful, self-propelled driverless transport robots are able to load movable shelves and carry them to the order picker. The new robot staff allows DB Schenker to increase efficiency by 60% and to reduce the warehouse area by 20% (Harttmann 2016).

Despite the automation, there is still demand for skilled workers. According to the logistics trade group Material Handling Industry (MHI), the logistics business in the USA will be looking to fill about 270,000 jobs per year, not only drivers, dispatchers, and clerks but also engineers, seasoned managers, marketers, data analysts, and human resources executives (Fisher 2014). New ways have to be found to attract talent and to encourage people, either college graduates or more mature job-seekers, to work in the supply chain industry.

New programs must encourage employees to stay longer in the job. In addition to their work, their experience and knowledge is highly valuable. By not encouraging employees to share their knowledge, companies in the Fortune 500 are losing an estimated combined $31.5 billion per year (Myers 2015). The private and public sectors share a responsibility to take advantage of human assets and develop programs to facilitate knowledge sharing between the different and diverse groups in the organizations.

3.1.4   Gender Diversity

Traditional gender roles are becoming a thing of the past. Today, married women are likely to out-earn their husbands and serve as the primary breadwinner for their family. Women contribute to the household income in 40% of American households with children under 18 (11% in 1960) indicated by a 2013 study by the Pew Research Center (Pew Research Center 2013). Fifty-seven percent of women are employed, according to U.S Department of Labor; in 2014, their most common jobs were secretaries and administrative assistants, elementary school and middle school teachers, and registered nurses. The number of women in professions such as architects and naval experts, production workers, machine operators, and geoscientists is also growing, professions which are usually male dominated (United States Department of Labor 2016).

This enables organizations to draw talent from among women in order to fill the skillset gap. But to do so, new ways to attract female workers might need to be found. An increasing number of work-from-home or shared part-time jobs could reconcile family and work life for both men and women.

How Gender Diversity Impacts Global Supply Chains
New demands lead to new offers. Working women are a consumer class. As a logical result, the supply chain needs to be designed for female service offers. While women increasingly work, deliveries at home are rising too. CSCOs are faced with the challenge of an increasing number of failed delivery attempts to private customers. As a consequence of couples working, the share of parcels successfully delivered on the first attempt is decreasing. This leads to increased costs and reduced customer satisfaction. The logistics service sector has embraced the trend by installing lockers and drop boxes. Another likely future service might be a higher number of female delivery personnel to respond to women's desire for security.

Supply chain managers and policymakers should heed women as employees, especially in less traditional segments of women labor. The working environment needs to be tailored to female needs. Ergonomic workplaces and tools, such as robotic exoskeletons that make lifting and holding of heavy things easier, have been developed in Japan in order to meet the needs of an aging labor force; in the USA, Harvard University is working on a similar project. Women can benefit from this development too.

The brand image of the supply chain sector needs to be lifted, not only to attract more women but also to raise the overall attractiveness of the industry. Human resources departments and managers able to meet female requirements will be among the winners of the gender race.

3.1.5   The New Consumer

Do consumption patterns promote technology, or does technology determine consumer behavior? Even though this question cannot be answered easily, one fact seems to be certain: in the digital age, consumption patterns have experienced a radical change. Around the world, product ranges are widening; offers are manifold; and delivery times are shorter than ever. The number of consumers and producers is growing. Due to the possibilities of the Internet and many commercial platforms, all can be connected. But with the number of connections, complexity along the supply and value chain is constantly rising.

Demand for organic products is growing. Whole Foods Market, a leading natural and organic food supermarket in the USA, released figures in 2010 that stated that "27% of adults say that natural and/or organic foods comprise more than one-quarter of their total food purchases." This number had increased from 20% in 2009 (Whole Foods Market 2010).

Not only in the West do consumers request sustainably grown and fairly traded produce, but also in the eastern nations, such as China, the number of people purchasing organic products is also increasing. This is mostly a result of growing middle class purchasing power. Another driver is the higher transparency along the supply and value chain unveiling major flaws, such as the "cadmium rice" incident in 2014, when heavy metal pollution of China's rice crops was proven, and the Chinese milk scandal in 2008. The need for safe products opened a new market for Chinese consumers heading to Hong Kong and Europe to buy milk powder for babies for resale in China (Song 2015).

Mass customization is another major trend.[3] On one hand, unique products are requested. On the other hand, demand for standardized, low-cost products is booming. Mass customization combines standard mass production with personalization and adaptability. In the process, mass-produced products become unique. Apart from the need of a factory capable of mass customization, the supply chain needs to enable or be part of the personalized service. Ohio-based detergent manufacturer, *ChemStation*, has been using a distribution system, which digitally connects 41 production units all across the country to combine its custom cleaning products to individual deliveries (Martin 2015).

Self-sufficient living is another emerging trend. Tips for urban homesteading can be found all over the Internet and in magazines. People are putting up solar panels and domestic cogeneration units to produce electricity. Flat roofs and balconies are turned into gardens; allotments are becoming popular in cities. In its wake, peer-to-peer businesses have become popular.

---

[3]Mass customization refers to the process of delivering wide-market goods and services that are modified to meet specific customer needs. It combines the flexibility and personalization of "custom-made" with the low unit costs associated with mass production. Another term is "made to order" or "built to order."

Above all else, the popularity of online shopping is increasing globally. Current e-commerce statistics state that 40% of worldwide Internet users have bought products or goods online. This amounts to more than 1 billion online buyers and is projected to continuously grow. In the USA alone, 75.8% of the population was shopping online in 2013. Driven by creative omni-channel offerings and increased cybersecurity efforts,[4] the market growth is projected to accelerate in most countries (A.T. Kearney 2015).

According to e-commerce Europe, China is the world's largest e-commerce market at $538 billion, followed by the USA at $483 billion. Asia Pacific is the largest region ($770 billion) with also the highest growth (44%) of the top five, followed by Europe ($567 billion, 14%), North America ($523 billion, 12%), Latin America ($37 billion, 18%), and Middle East and North Africa ($21 billion, 22%). The remaining countries account for $25 billion with 24% growth.

Border-crossing e-commerce is increasing too. Western Europe is one of the largest markets in the world for cross-border trade. In the USA, 34.1 million people are online cross-border shoppers. In 2013, US e-commerce customers mostly shopped from the UK (49%), China (39%), Canada (34%), Hong Kong (20%), and Australia (18%) (thepaypers.com 2016). With its project Supply Chain by Amazon, aimed at entering the global logistics market, the company is challenging China's Alibaba Group for global e-commerce leadership. At stake is a projected $1 trillion in cross-border e-commerce revenue by 2020, experts say.

Things are sought, offered, and purchased in increasingly shorter intervals. If in the past, people bought their clothes with the change of seasons, consumers nowadays expect new offers or new models with much higher frequency. Once a new smartphone is on the market, the old one goes, no matter its condition. This places major strain on the environment.

Consumers are also starting to become their own producers. With websites such as Etsy and easy to create customized products, such as on-demand printing for a wide range of less complex goods, from cups to shirts to books, a new class of entrepreneurs emerge, the "prosumers," consumers who become involved with designing or customizing products for their own and other people's needs. They usually produce at home.

China is today's most important consumer market. Although <2% of Chinese workers earn enough to pay income tax, numbers are big in China. According to Goldman Sachs, the working population amounts to 770.4 million people. In comparison, 146 million people are working in the USA. The annual income per capita of the middle class in China amounts to $11,733. Today, the middle class spends the majority of their income on clothes and food. In the future, higher spending on trips abroad, technology and media, use of mobile data (costs are currently high, keeping data usage low), and healthcare is expected. Consumption

---

[4]Omni-channel retailing is a business model that uses a variety of channels such as retail stores, online stores, mobile stores, mobile app stores, telephone sales, and any other method of transacting with a customer including research before purchase.

has been driving growth for the supply chain and transport industries. Just for e-commerce, shipping grew ten times from 2006 to 2014 (Goldman Sachs 2016).

How the New Consumer Impacts Global Supply Chains

The extended range and diversity of products, paired with increasingly higher frequency of purchases and shorter use cycles, will drive a continuous increase in complexity along the supply and value chain. Rising global consumption and purchasing power calls for dense global supply chain networks. The Asia Pacific region, which was home to 18% of the world's middle class in 2009, will have 66% of this important consumer class by 2030, according to the Development Center of the Organization for Economic Cooperation and Development (Escobar 2015). This drives the need for significantly reshaping global supply chains.

Consumers now expect their purchase to be available to them instantly, or at least in a matter of hours and sometimes days, leading to the demand of a swift supply chain. What transportation services will be required in the world of 3D printing? There were no more planning, sourcing, making, and delivering. Instead the supply chain will need to react to the prosumers' orders and take care of production, delivery, and possibly recycling. On the other side, retailers and delivery companies need to accelerate shipments. Not being able to step up to the high customers' expectations might diminish reputation. Some retailers might fall back on private or dedicated fleets in order to react faster and offer shorter lead and delivery times.

Self-sufficient living is based on the peer-to-peer concept. The logical extension of the peer-to-peer economy is the peer-to-peer delivery network. The sharing economy, also known as the peer economy, collaborative consumption, or "shareconomy," is becoming a popular model – from eBay to Airbnb and Uber. With the help of the Internet or apps, anyone can sell or borrow things, rent out a flat, or offer a ride in a car.

Some experts have made out another new trend developing: platform coops. In the case of transportation networks, such as Lyft and Uber, the cooperative business would be governed by drivers. In contrast to tech start-ups that are all about revenue, the central idea behind platform cooperativism is that those who create the most value should own and control the platforms. According to *The Economist*, there are already a few platform coops out there, trying to stand up to their successful tech counterparts. There are three common developmental patterns to be found. One of the examples is *Loconomics*, the cooperative version of task marketplace *TaskRabbit*, a legally defined cooperative version of sharing economy platforms. One of the most successful experiments is *Enspiral Network*, a New Zealand-based co-working community plus digital collective that allows hundreds of freelancers and social enterprises to work together for mutual benefit. Last but not least, *LaZooz* is the blockchain[5] version of transportation networks where

---

[5]Blockchain is a database that maintains a growing list of data records secured from tampering and revision.

drivers earn their share by giving rides (Gorenflo 2016), similar to *Arcade City*, which was launched by former Uber drivers (Valenzuela 2016).

Parcel pickups at prosumers' homes are hard to handle. Delivery companies have come up with new services, ranging from drop-off and pickup boxes to call-a-parcel carrier. Will the prosumer also become part of a transport network in the sharing economy? "Crowd-shipping" might become more influential in regard to shipping and consumption. The sharing economy enables new ways of transportation by connecting those who need parcel deliveries with those who are already on the road. The main driver is the desire to find new ways of overcoming the traditional problems of "last mile" delivery. Like *BlaBlaCar*, the city-to-city car-sharing service, crowd-shipping offers a means of better using underutilized cars through picking up and dropping off parcels along the routes people might be taking anyway. Some platforms cater more for professional couriers: for example, it is estimated that at *Zipments*, 95% of couriers are professionals with more than 4 years of experience. Others such as *Rideshare*, *MyWay*, and *Shippies* rely on everyday people (Lehmacher 2015).

The level and detail of information available to consumers is growing at a steady pace. At the same time, compliance to societal and ecological standards is becoming more important. Product scandals are not without effects for the producers, and shipping companies can be affected too. *FedEx*, for example, risked fines of up to $1.6 billion for shipping illegally sold drugs (Chumley 2014). Following the China scandal, milk powder became a significant part of the business for some global transport operators.

Mass customization adds another layer of complexity to the supply and value chain and drives new supply and value chain designs. In a survey by GT Nexus on top supply chain issues, more than one-quarter of respondents (27%) said keeping up with customer demands was their company's number one challenge (GT Nexus 2016). Mass customization requires not only very specific value chains, but part of the uniqueness itself can be the special service offered at delivery. Forward-looking supply chain design takes this into account by providing product and other information to the public and actively involving business customers and consumers – in the return process of used-up parts and products at the end of the use cycle, for example.

3.1.6 Urbanization

Eighty-three percent of the online shoppers in the USA live in cities (thepaypers. com 2016). Migration from rural to urban areas in many parts of the world will impact consumption patterns and, thus, the supply chain. The percentage of population that resides in an urban setting surpassed the 50% mark sometime between the year 2000 and 2004. By 2025, China's urbanization rate is projected to be around 60%, India is likely to reach 36%, and the USA will be slightly shy of 90%, compared to 82% in 2014, according to the United Nations' 2014 World Urbanization Prospect.

China in particular has experienced a major shift towards urbanization; in 1990, the urbanization rate was 26%. The economic development in the eastern part of the country resulted in an urban and industrial corridor stretching from Harbin in the northeast through the Beijing area and south to China's metropolitan complex of Shanghai, China's largest city. Well-paying jobs provided people with a better standard of living, drawing even more people into the area. In 2013, the workforce is larger than those of the USA and Europe combined. China's middle class is growing, and rising incomes are creating a new class of consumers.

The increasing number of urban dwellers will affect traffic, passenger, and goods mobility. In the past, infrastructure plans were rarely able to keep up with the urban growth pace. As a result, mobility and hence productivity within urban areas have been compromised. Ambulances and fire trucks regularly get stuck in traffic, and parcel carriers are often not able to deliver on time.

According to the annual Traffic Index by TomTom, the most congested cities in the world are Istanbul in Turkey, Mexico City in the United States of Mexico, Rio de Janeiro in Brazil, Moscow in Russia, and Salvador in Brazil (Kirkpatrick 2015). In the USA, an annual Traffic Scorecard, compiled by Texas A&M Transportation Institute (TTI), places Washington, DC, the larger Los Angeles area, and San Francisco/Oakland as the top three most congested cities in the country, with an average of 82, 80, and 78 h, respectively, spent in traffic per capita annually. Another more vivid example is a 20 mile stretch of the Cross Bronx Expressway in New York which is determined to be the most congested traffic corridor in the USA, taking 54 min to traverse at an average speed of just 13 mph.

Along with delays, citizens and governments have to deal with increasing costs and environmental pollution. By 2020, TTI projects that the total nationwide delay time will grow to 8.3 billion hours and congestion will cost $192 billion. Fuel wasted increased by 520% in 2015, from just shy of 0.5 billion gallons in 1982 to 3.1 billion gallons (Inrix 2015).

Growing agglomerations lead to rising demand on living space, schools, clinics, water, energy, and recreational programs; emissions increase too. The "United Nations Environment Programme" estimates that cities are already responsible for 75% of global CO_2 emissions, with transport and buildings being among the largest contributors (UNEP 2016). In 2011, cities contributed 70% of global GHG emissions while covering only 2% of global land mass.

How Urbanization Impacts Global Supply Chains
Cities are looking for solutions to avoid infrastructure investments and reduce energy use and emissions while still ensuring the fluid movement of goods. In 2013, *Amazon* announced it would test delivering packages of five pounds or less by drones. In China, *SF Express* has launched drones to deliver packages to remote areas. *Swiss Post*, *Swiss WorldCargo*, and *Matternet* are jointly testing the commercial use of logistics drones in Switzerland. *Maersk* is experimenting with drones for deliveries to vessels, a system that could eventually replace barges. The test drones can carry up to 1.2 kg. Maersk Tankers estimates that using drones could result in potential savings of $3000–9000 per vessel per year; the costs for a barge

are on average $1000 (Cage 2016). Network-owned drone services, in the sense of a blockchain, where the units could charge themselves independently, might help supply chain service providers to increase fluidity and reduce delivery costs even further.

According to a market research report, the unmanned aerial vehicle (UAV), aka drones, market is estimated to be worth $5.59 billion by 2020. A major growth is expected to happen in the commercial sector. In the Americas, the report estimates the market could grow almost 38% between 2015 and 2020 (Markets and Markets 2016). Drones might not represent the ideal solution for city deliveries but for suburban and remote rural areas.

Tunnel or pipe systems, under or above the ground, are other ideas to improve the flow of goods in urban areas and between cities. Switzerland is assessing the concept of a "nationwide, underground, autonomous transport network." The system comprises "a three-lane network hosting electric, drone delivery capsules." The rail-mounted vehicles are designed to load two standard pallets and will travel on maglev rail at 30 km/h (van Marle 2016). Elon Musk's *Hyperloop* concept suggests that the distance between Los Angeles and San Francisco can be traveled in 35 min. At that speed, the importance of the global web of cities becomes even more evident.

In London, autonomous ground vehicles, which are navigated to the customer's address, had a test-run in March 2016. The rolling robot, invented by Skype cofounders Ahti Heinla and Janus Friis, could help small business owners to serve local customers through automated goods deliveries and will be tested in New York next (Woollaston 2016). In Germany, *Deutsche Post DHL* already uses electric vehicles to deliver mail and parcel.

Clean goods mobility solutions enable delivery companies to enter cities when traffic regulations prohibit vehicles. Banning vehicles is probably not a workable solution, as city dwellers and businesses rely on transport companies for both goods supply and waste management.

A mobility concept for mountainous cities is urban ropeways as introduced by Austria-based manufacturer *Doppelmayr Seilbahnen*. The company constructs the "future of urban mobility above our heads." It has established the eco-friendly, mostly barrier-free ropeway cars in numerous cities such as La Paz in Bolivia, Caracas in Venezuela, or Portland in Oregon.

Spotlight: 100 Resilient Cities

Since 2013, the Rockefeller Foundation has been working on making cities around the world more resilient to the physical, social, and economic challenges they face. An initiative called 100 Resilient Cities (100RC) – pioneered by the Rockefeller Foundation – is providing cities that join the network with the resources necessary to develop a roadmap to resilience.

(continued)

Urban resilience is defined by seven characteristics that allow cities to "withstand, respond to, and adapt more readily to shocks and stresses." They have to be reflective, resourceful, flexible, and robust; redundant in the sense of creating spare capacity to handle disruptions; integrated in terms of bringing together various systems and institutions; and lastly, inclusive, for example, by integrating multiple stakeholders into the decision-making process. In the first year, 32 cities were awarded with a membership; in 2014, 35 cities joined. Among them are Amman, Jordan; Barcelona, Spain; Bangkok, Thailand; Christchurch, New Zealand; Melbourne, Australia; London, England; Montreal, Canada; Porto Alegre, Brazil; Singapore, Singapore; and New York, NY.[6]

3.1.7 Mobility in the Twenty-First Century

Infrastructure is simply not able to deal with today's heavy traffic. According to the American Moving & Storage Association (AMSA), the condition of U.S roads and bridges is deteriorating. They point out that 33% of the roads on the National Highway System are in poor or mediocre condition and 26% of America's bridges are structurally deficient or functionally obsolete (AMAS 2016). There are truck tolls in place in various States, which are increasing in fees and in number. In December 2015, US President Obama signed a 5-year, $305 billion highway bill to get infrastructure projects underway.

According to US publisher, Ward's, the USA operates the world's largest fleet of motor vehicles, with 239.8 million by 2010 (Wikipedia 2016). Also, vehicle ownership per capita in the USA is the highest in the world with 769 vehicles in operation per 1000 inhabitants or the ratio of 1:1.3 vehicles to people. In the USA, the number of newly registered cars remained relatively consistent over the past 5 years: 255,876 (2013) and 255,212 (2009) (Statista 2016). In Europe, the overall passenger car fleet in almost all of the EU member states has grown over the last 5 years. Internationally, total vehicle sales were at about 77 million in 2012, a 6% increase from 2011. This included about 50 million cars and 27 million commercial vehicles. In 2012, the fastest growing markets were Thailand (+80% from 2011 to 2012), Japan (+28%), and Indonesia (+25). Since 2007, vehicle sales have about doubled in these countries along with China, where 19 million vehicles sold in 2012 (ICCT 2013).

According to Eurostat, diesel-driven cars still top the overall number of new registrations. In 2014, 58% of the new passenger cars were powered by diesel engines, 38% by petrol engines. Four percent of the passenger cars run on alternative fuel engines such as liquefied petroleum gas (LPG), natural gas, electricity, or

---

[6]More information at http://www.100resilientcities.org

other alternative fuels (Eurostat 2016). In the USA, the number of alternative fueled and hybrid cars added up to 1,170,164 in 2014. In the same period, 1,684,463 commercial vehicles were running on alternative fuel, ranging from minivans to heavy-duty trucks (US Energy Information Administration 2016).

Policymakers are looking for ways out of the mobility dilemma, in both the freight and the passenger traffic sector. Combining different modes of transport, such as rail and road, improves the efficiency of transportation. Intermodal transport solutions are becoming increasingly popular. Self-driving cars are largely seen as a last mile solution, complementing public transport.

By designing self-driving cars, *Google* has demonstrated the tight link between the Internet and mobility. Major car manufacturers are seizing the opportunity and developing self-driving cars, including *Volvo*. In 2015, the Swedish company went public with the Drive Me project, a solution that aims to make it possible to integrate self-driving cars into real traffic (VOLVO 2015). Policymakers are already responding. The British Department of Transport released a set of rules for testing driverless cars on public roads in 2015; human drivers are still required to take over when needed (Walton 2015).

The German vehicle manufacturer *Daimler AG* sent the world's first self-driving truck on the highway. The Freightliner Inspiration received official licensing to use the public highways in the state of Nevada in 2015 (Lavars 2015). The benefits of self-driven vehicles are manifold: due to their utilization of telematics,[7] they cause less accidents and manage driver behavior; they use less fuel, and thus reduce emissions and costs. Platooning, the use of digitally connected convoys driving closely behind each other, was tested by *Peloton Technology, Meritor WABCO*, and *DENSO* in Detroit with two tractor-trailers in 2014; *Volvo* and *Peterbilt* are also developing and testing platooning solutions (Lockwood 2016). The system is ranked as a "tool for solving transportation problems" by the U.S Department of Transportation, and it is putting R&D into this topic (Dopart 2015). Benefits of tractor-trailer convoys are fuel efficiency, less required space, and higher safety.

In Ann Arbor, Michigan, the University of Michigan has built a simulated urban zone setup to test self-driving and connected cars. This Public-Private Partnership (PPP) program operates under the name Mobility Transformation Center (MTC) and includes *Ford, GM, Honda, Toyota*, and *Nissan*, as well as suppliers *Delphi, DENSO, Navistar, Qualcomm, Verizon, Xerox*, and Michigan Department of Transportation. The 32 acre patch on the University's North Campus consists of two- and four-lane roads with concrete, asphalt, brick, and dirt surfaces, adjustable street lighting, sidewalks and fire hydrants, crosswalks, curb cuts, bike lanes, a railroad crossing, a pair of roundabouts, highway entrance ramps, and lots and lots of

---

[7]Telematics – umbrella term that encompasses telecommunications, vehicular technologies, road transportation, road safety, electrical engineering (sensors, instrumentation, wireless communications, etc.), and computer science (multimedia, Internet, etc.). Most commonly known telematics solution is the navigational system.

parking spaces. Buildings can be moved depending on the test design (Motavalli 2015).

The German Car Industry Association [German: Verband der Automobilindustrie (VDA)] and the Federal Institute for Highways (BASt) have defined three levels of autonomous driving:

- Semiautonomous vehicles: The driver is constantly monitoring automated functions.
- Highly autonomous vehicles: Activities not connected to driving are possible to some extent, since the driver does not need to monitor driving at a constant level. If the system reaches its limits in regard to automation, the driver will take over in time to prevent incidents.
- Fully autonomous vehicles: The vehicle's system is able to totally integrate into traffic without intervention of the driver. Thus, drivers are able to attend to operations not connected to driving such as checking freight documents.

A first step towards the latter was taken in February 2015, when the U.S National Highway Traffic Safety Administration (NHTSA) agreed on considering "driver" in a broader sense, when dealing with Google's submitted proposed design to allow the Google car on American roads. According to the news agency Reuters, the NHTSA stated: "We agree with Google that its (self-driving car) will not have a 'driver' in the traditional sense that vehicles have had drivers during the last more than one hundred years." NHTSA officials continue: "The next question is whether and how Google could certify that the (self-driving system) meets a standard developed and designed to apply to a vehicle with a human driver" (Shepardson and Lienert 2016).

Intelligent Transport Systems (ITS) need to be established. These advanced applications aim to provide data to manage traffic flows to a better degree, make roads safer, and enable the various users to be better informed about road conditions. An ITS provides data for drivers to help them find available parking, making compliance with parking regulations easier for professional drivers.

In the USA, the Department of Transportation has formulated the ITS Strategic Plan 2015–2019. The framework is built around two key ITS program priorities: first, the realization of connected vehicle implementation, which builds on the substantial progress made in recent years around design, testing, and planning for deployment of connected vehicles across the nation, and, second, advancing automation by way of research, development, and adoption of automation-related technologies as those emerge (United States Department of Transportation 2015). In the EU, a legal framework for ITS was adopted in 2010 (Directive 2010/40/EU), which aims to accelerate the deployment of these innovative transport technologies across Europe.

The Hamburg Port Authority (HPA) in Germany has introduced smartPort Logistics. The system navigates trucks as these enter the harbor area towards their intended loading or unloading area. In case of delays, the system informs the driver whether or not the time can be used for additional pickups or drop-offs.

Using smartPort Logistics, HPA strives to increase efficiency and reduce congestion (Hamburg Port Authority 2016). Autonomous vehicles would be able to operate in an interconnected world efficiently and could contribute to smoother traffic and cleaner mobility. Even though autonomous driving is already a technological reality, legal frameworks and measures in regard to insurances are still missing.

How Twenty-First-Century Mobility Impacts Global Supply Chains
Population growth and climate change call for smart mobility concepts. The new models require collaboration between different industries. Supply chain service providers and transport companies must collaborate with vehicle manufacturers and technology companies. The aim is to create cost-effective, fluid, and sustainable concepts. Policymakers need to be continuously involved in development thinking and process.

The supply chain will be increasingly shaped by digitalization, making way for innovative process design and business models. Customers are currently able to track and trace their orders and get an estimated time of arrival (ETA) with constant updates. Unsuccessful deliveries could be a thing of the past if customers could simply redirect their shipment to another address, drop-off box, shops, etc. Customers already receive a notification when goods are delivered at a drop-off box along with a code that allows them to retrieve the order.

Transport network operators need to integrate the new possibilities or interface with new operators offering transport capacity. Innovative technologies are going to be utilized all along the supply chain, not only as shop-and-ship services or citizens performing deliveries but also on freight exchange platforms such as *TimoCom*, *uShip*, or *LoadMe* that reduce the number of trips and dead mileage. With systems that derive their design from social media platforms, customers can easily offer and accept services and freight.

3.1.8 Digital Culture

In 2015, the number of worldwide email users was nearly 2.6 billion. By the end of 2019, the number will have increased to over 2.9 billion that's over one-third of the worldwide population. In 2015, the number of emails sent and received per day total over 205 billion. This figure is expected to grow at an average annual rate of 3% over the next 4 years, reaching over 246 billion by the end of 2019. A recent study by the Palo Alto-based *Radicati Group*, a technology market research firm, projects continued strong use of email in the business world, as well as among consumers. The amount of consumer email will continue to grow, mainly due to its use for notifications, e.g., for online sales, rather than simply as an interpersonal communication tool, the report predicts (The Radicati Group Inc. 2015).

Emails are not solely being sent or received via computer either; tablets, smartphones, smartwatches, and smart TVs all handle digital mail. The Internet is growing to be the universal tool of communication, collaboration, and creation, connecting business and domestic life. Smartphones act as a control center for energy supply, banking, mobility, and health; sensors and satellite technology such as GPS and RFID distribute data over the Internet; purchases, orders, and reservations are made with one click. Apps form the interface between the social and semantic web.[8]

Computerization and digitization are not without challenges, though. According to a 2013 study by Oxford Martin School, 47% of jobs in the USA could be computerized within one or two decades. This might not only affect manual labor jobs but also many cognitive tasks, due to the availability of big data. The study suggests two waves of computerization, with the first substituting computers for people in logistics, transportation, and administrative and office support and the second affecting cognitive jobs, depending on how well engineers crack computing problems associated with human perception and creative and social intelligence (Frey and Osborn 2015).

On the other hand, computerization fosters entrepreneurship. A survey conducted in 2015 by the World Economic Forum suggests that entrepreneurship could help bring youth unemployment down. 65% of the 2800 young people surveyed around the world want to run their own business at some point in their careers. These ambitions are particularly strong in large emerging markets, led by Mexico (91%) and China (89%) (Pinelli 2015).

With digitization, everyone participates in cultural, economic, and social life; families are able to communicate with their loved ones across oceans. Possibilities within any business sector seem endless. Almost every business transaction can be traced. This allows improvement in quality of life and operational business processes. Digitization can reduce corruption through a higher level of business transparency and e-government; uncertainties and misconduct can be diminished. Things not entered in the system could be considered illicit and trigger scrutiny and inspection.

By way of increasing interconnectivity, boundaries between producer and customer, work and leisure, living and working space are blurring. Living quarters of the future will be multifunctional, allowing employers and employees to efficiently work from home while being part of both family and work teams. Meetings across long distances and time zones, from offices, home offices, or hotel rooms, using video and phone conferencing are already a common practice. Expenditure of time and expenses, along with CO_2 emission, is reduced – because travels become obsolete to a certain extent. There are even projects that strive to bring the workspace onto the road to increase productivity and mobility (Lanks 2015).

The sharing economy may terminate the age in which capacity and knowledge is power and thus an advantage in regard to career and competition. Modern times are

---

[8]The Semantic Web is an extension of the Web through standards by the World Wide Web Consortium (W3C). According to the W3C, "The Semantic Web provides a common framework that allows data to be shared and reused across application, enterprise, and community boundaries." (Source: Wikipedia)

more about identification and intelligent combination of competences over the Internet – mainly the ability to communicate and collaborate to create value-added applications.

How the Digital Culture Impacts Global Supply Chains
Interconnected economic and social networks have become reality, impacting the planning and management of the supply chain. Home offices and virtual teams help to meet the expectations of modern lifestyle and attract talent. Platforms help to plan and manage the ecosystem and drive entrepreneurship. Producer and consumer interact digitally, taking on a more active part in designing and controlling the global supply chain. Through smart devices and apps, they influence ordering, delivery routing, and time and place of delivery or pickup of shipments. Consumers become partners in supply chain design and take over tasks.

Enterprise and society, work, and leisure will continue to become increasingly digital, opening up new opportunities of collaboration. The supply chain of the future will deal with digital customers, employees, and partners. CSCOs might soon become digital system designers.

3.1.9 Ubiquitous Intelligence

The Internet of Things, cloud computing, robotics, and artificial intelligence (AI) are closely linked. New technology development has unprecedented momentum. One example is minicomputing, intelligent devices connected to the Internet: IBM predicts that by 2020, 50 billion devices will be connected to the Internet, a tenfold increase from the number in use today (IBM 2015). The Internet of Things (IoT) is the most important technology trend for the supply and value chain. This infrastructure of connected intelligent devices and assets will change the design, planning, and management of the flow of goods and transform the linear chain into an orbit of unlimited possibilities.

Cloud computing, with its enormous storage capacity, is becoming a standard across the world of business and works in complement to the Internet of Things. Comprehensive interfaces and smart systems, backed by cloud storage, enable authorized users to access data, materials, information, documents, etc. from anywhere, anytime. Substantial investments are no longer necessary from individual companies to build this infrastructure, making the cloud a suitable tool that offers large storage capacity even to small businesses.

Robots assume jobs in various parts of the supply chain such as in warehousing, in moving freight, in delivering parcels, etc. Robocalls take care of customers or write newsletters to keep the customer base informed about changes, offers, and the latest developments. Artificial intelligence (AI) robots or computers can solve problems through comparing gathered data to find the best possible action; learning robots recognize whether or not actions achieve the expected result and store information to apply or avoid the actions in a future similar situation.

Small programs called apps are produced in large numbers and for any situation or occasion. They are starting to encroach on and shape our entire personal and professional lives. Apps will see an even greater uptick as the IoT leads to more use cases and new business models. Sensors are placed in many products and will only increase in utilization in the near future. The IoT will change the way we use things. Intelligent walls and dashboards with sensors and microcomputers provide information. Interconnected devices establish a world of machine-to-machine communication and open up new experiences.

Wearable technology, including clothing and accessories incorporating computers and electronics, will also offer solutions to many sectors along the supply chain. Devices using augmented reality have appeared in the sector through 3D smart glasses by *Atheer*.[9] Equipped with smart glasses and linked to the platform, workers can optimize the workflow related to every pallet or shipment build, from picking order to item stacking, beyond past warehousing experience. At *DHL*, the augmented reality application in warehouses improved picking process by 25% (DHL 2015). Other pioneers in the field include Fujitsu *FEELthym*, a wearable device that detects drowsiness or fatigue of long-haul truck, bus, and taxi drivers; the *Pebble* Smartwatch by W&H Systems' Shiraz allows warehouse managers to gather critical performance indicators to better manage their distribution and fulfillment operations.

The key benefits of wearable technology are largely seen in making processes faster and more accurate, because it automatically provides the worker with information. The "thinking" is done in the systems and supported by artificial intelligence which also helps the system to learn. No one needs to look for and work through paperwork or search for parts in the warehouse. The process information generated can be used to detect and eradicate inefficiencies. Last but not least, safety and health issues are addressed. Fitbit Activity Tracker by *Fitbit Inc.*, for example, gathers health information and provides recommendations on when to take a break. Reduction in accidents and injuries is one of the benefits; another is higher employee satisfaction. Smart concepts will flood our lives; some of them will stay because they are truly useful. Some will be short-lived.

How Ubiquitous Intelligence Impacts Global Supply Chains

Today's supply chain without information technology is unthinkable. The success of tracking and tracing is the early evidence. The opportunities technology has to offer will be the basis for the next step of evolution: smart devices in operation, or integrated in almost everything, will support new practices and business models.

The Internet of Things will play a major part in developing flexible, effective, and resilient supply chains. Sensors and RFID tags transmit data that is continuously analyzed. Through enrichment with further data, such as planning

---

[9]Augmented reality is a live direct or indirect view of a physical, real-world environment whose elements are augmented (or supplemented) by computer-generated sensory input such as sound, video, graphics, or GPS data. (Source: Wikipedia)

information and contingency plans, this might lead to a renaissance of the centrally managed supply chain on a global level. Shipping routes could be centrally coordinated. Wrong loads, disruptions during shipment, or delays due to customs procedures could be recognized immediately and fixed accordingly. With the aid of the IoT, CSCOs will be empowered to handle the increasingly complex supply chain. Thanks to information transmitted by ICT systems, along with options to intervene, potential cost reductions could emerge. Also, risks of delays and supply shortage could be reduced. The systems are valued decision-making tools, which could be programmed to make decisions by themselves (IBM 2010). The IoT combined with wearables leads to augmented reality, which will enable employees in warehouses and vehicles to work more effectively and prevent accidents or human errors. Robots will increase precision and efficiency in handling.

In the IoT, information on flows and conditions of goods is accessible anytime and anywhere: a device, even though not in use, is still able to transmit its position and status. An example is the RFID chip that sends out data to designated recipients 24/7, which allows constant monitoring of the movement of goods along the chain, optimizing planning, quality, and costs along the way. Other devices only start working when needed, which is highly energy efficient. Viable solutions need to turn fast and target-orientated data into competitive advantages and new business models, with better customer experience, lower cost, and lower rate of error to drive loyalty and sales.

This does not absolve CSCOs from the responsibility to consciously evaluate events along the supply chain and to compare set goals with actual values. Humans should still be the final source of decision-making and control. IT systems do not make collaboration of all entities from suppliers and manufacturers to service providers, business customers, and consumers unnecessary.

Ecosystems will become more intelligent, creating new opportunities and challenges along the supply and value chain – at touchpoints, in operations and planning, monitoring, and management. Almost unlimited traceability combined with analytic tools produces information that makes way for new ways of operating and new services in areas such as maintenance, recycling, and customer care.

The open flow of massive amounts of data will be a powerful tool, but it is not without its own drawbacks. Cybersecurity will be one of the key challenges for the future, not only in ICT but for business, government, and society in general. CSCs need to ensure that the supply chain cannot be hacked. Blockchain could be a turning point for cybersecurity. Blockchain is a database that maintains large amounts of records, secured from tampering and revision by being distributed across many users. It was developed to track bitcoin, the Internet-based currency, but has natural applications in supply chain as well. It could be used to record all paperwork related to a shipment, each time it changes hands and every time a related payment is made. Conflict minerals or poached wildlife products could be identified at their source. The updated information would be available to every user in the chain almost immediately, creating unparalleled visibility, while the decentralized and constantly rewritten record makes it harder to tamper with the information, providing security and certainty (Vorabutra 2016).

IBM has launched a blockchain service that lets companies build their own blockchains in IBM's secure cloud. It is aimed at supply chain companies that need to track high-value items through complex movements. One of the inaugural clients is *Everledger*, which tracks the provenance of diamonds to identify stones mined in conflict regions or those known for forced labor (Nash 2016). Mining giant *BHP Billiton* has announced that it will use blockchain to track the movement of samples for testing (Rizzo 2016).

Several companies, including *JP Morgan Chase*, are developing blockchain platforms based on *Ethereum*. Ethereum is a major distributed-ledger project created by 21-year-old Canadian programmer Vitalik Buterin. It can process more data than bitcoin's ledger and comes with the Solidity programming language, which allows users to write smart contracts. Smart contracts can do things such as create invoices that pay themselves when a shipment arrives or send owner's dividends if profits reach a predetermined level (The Economist 2015).

3.1.10 Bionics: Learning from Nature

The modern supply chain is an interlaced ecosystem which mimics complex and efficient systems in nature. Bionics increasingly find their way into research institutions, business operations, and fabrication.[10] They inspire, give input, and deliver solutions in the design of products and services or organizational structures. For example, organizations can benefit from so-called swarm intelligence, the collective behavior in self-organized systems. By collaborating, organizations and ecosystems are able to react more efficiently to disruptions and also combine knowledge, experiences, resources, and machinery in order to get work done faster and more ecologically. Applied swarm intelligence in robotics, so-called swarm robotics, is the basis of some warehouse automation systems.

Nature has been the inspiration for, and designer of, many useful inventions, such as super strong adhesives based on the mechanics of gecko feet and self-cleaning, water-repellent lacquer mimicking the lotus leaf. Distance warning devices on cars and parking sensors copy the sonar systems bats use to maneuver, while the marine industry designs vessels by adapting the dolphin's body shape in order to create an optimal flow resistance. The aviation industry took eagles and condors as role models for designing wings that reduced wakes using integrated winglets or wingtips, which improved the efficiency of fixed-wing aircraft and reduced noise. *Airbus*, the aircraft manufacturer, has designed a Concept Cabin that includes bionic structures, biopolymer membranes, and composite and morphing materials (Airbus 2016). Artificial neural networks (ANN), used in AI, and in particular in deep learning, enable machines to learn complex operations and thus improve their own abilities. ANNs are applied to vehicle control, radar

---

[10]Bionics is the application of biological systems found in nature to understand and design business ecosystems and modern technology.

systems, facial recognition, and automated trading systems. ANNs are also used in clinical diagnostics.

Just as they do in nature, materials and compounds should create an endless cycle that will produce value with every utilization. This concept is known as cradle to cradle, a biomimetic approach to the design of products and systems. Only 15% of used cell phones are estimated to be collected. Waste Electrical and Electronic Equipment (WEEE) volumes in Europe for IT and telecommunications equipment are estimated at 825,000 t/year. Since the waste contains valuable materials such as gold, silver, and rare earth metals, the uncollected devices represent a material loss of up to $500 million annually (Ellen MacArthur Foundation 2012). The growing mountains of waste created by single-use items and the packaging required by increasing e-commerce urgently require a more sustainable solution.

How Bionics Impacts Global Supply Chains

One can identify biological value chains in nature, which serve as models for businesses to create an efficient and innovative supply chain. Whether we steer the chain deliberately in a circular way or not, we are living in a circular world. We must bring materials back into the cycle. We either create products well prepared for further use or dispose of out-of-use products as hardly digestible waste in landfills and oceans. In light of the negative effects of the take-make-waste society – e.g., reduced but still existing extreme poverty, global population growth, scarcity of resources, and climate change – the make-use-return circular economy (CE) is possibly the only route to escape the self-destruction trap and make more responsible use of the planet's resources and the nature around us. The backbone of the CE is the circular supply chain. CE and circular supply chain are currently lagging in implementation, though, because CE is widely seen as a technical concept rather than a holistic way of thinking and acting.

3.1.11 Resource Scarcity and Energy Shifts

According to the International Energy Agency (IEA), energy use worldwide is set to grow one-third by 2040. The primary drivers are India, China, Africa, the Middle East, and Southeast Asia, due to demographic and economic developments. China is predicted to overtake the USA as the biggest consumer of oil and to have a larger gas market than the EU. India and China are expected to account for half the growth in global energy demand. The share of nonfossil fuels in the global mix will also increase from 19% today to 25% in 2040 (International Energy Agency 2015).

The global transportation ecosystem is 96% dependent on oil, with aviation above 99%, and approximately 60% of oil is used for transport. Global energy demand for transportation is projected to rise by 40% from 2010 to 2040. However, since vehicles are predicted to be much more energy efficient, the demand for fuel will not increase significantly, states the "The Outlook for Energy: A View to 2040" by ExxonMobil (ExxonMobile 2016). According to the paper, by 2040, energy for moving freight and people by aircraft, marine vessels, and trains will be just shy of

30% (today: 20%); energy demand for aviation, marine, and rail will grow by an average of about 75%. ExxonMobil expects that most of the growing demand will be met by oil.

The global top three countries with proven oil reserves are Venezuela (298.3 billion barrels), Saudi Arabia (267 billion barrels), and Canada (172.9 billion barrels); the USA comes in ninth, accounting for 48.5 billion barrels (Word Atlas 2016). At current rates of production, the remaining lifetime of the various energy sources will be longest for coal (119 years), followed by gas (63 years) and oil (46 years) (Roland Berger 2011).

Companies are not only finding new ways to do away with fossil-based energy resources, but R&D are also continuously driving more energy efficient devices such as vehicle engines or appliances. The IEA "Medium-Term Renewable Energy Market Report 2015" suggests the share of renewable energy will rise to over 26% by 2020, from 22% in 2013 (IEA 2015). The agency expects the role of renewable energy to increase particularly in the power generation, heating and cooling, and transport sectors. A recent paper, published by the World Economic Forum, states that the public sector needs to become more active in regards to R&D of renewable energy. The Advanced Research Projects Agency-Energy (ARPA-E) had a budget of $280 million in 2015, not even one-tenth of the Defense Advanced Research Projects Agency (DARPA) budget. In 1981, energy accounted for 11% of the total U.S public R&D budget. Today, it accounts for just 4%. The public sector bodies that engage themselves in promoting green-energy technologies are state development banks, such as Germany's KfW, the China Development Bank, the European Investment Bank, and Brazil's BNDES, four of the top ten investors in renewable energy, amounting to 15% of total asset finance (Mazzucato 2015).

The wind energy sector has experienced significant global growth in the past years. At the end of 2015, the total global capacity reached 432,419 MW (Cumulative Installed Wind Power), representing growth of 17%. In China alone, the market grew by 22% resulting in a total installed capacity of 145.1 GW; U.S market reached 8598 MW or 8.6 GW (Annual Wind Power Generation); Germany led the performance in Europe with 6 GW of new installations, including 2.3 GW offshore (Global Wind Energy Council 2016). Photovoltaic systems are also on the rise. The top installers of 2014 were China, followed by Japan and the USA, while the UK emerged as European leader ahead of Germany and France (Solar Energy Industries Association 2016).

Water is becoming scarce. Assuming average economic growth without efficiency gains, water demand is projected to increase by 55% globally between 2000 and 2050. The increase in demand will mainly come from manufacturing (+400%), electricity (+140%), and domestic use (+130%) (Leflaive 2012). The total water demand will increase faster (by 18%) in the developing nations than in the developed countries. Even though developed countries will also need more water by 2030 (+40%), their share will decrease to 24% of the world's total water utilization (today: 27%). Water withdrawals from industry will grow the fastest over the next 20 years, rising from 16% today to 22% in 2030. Municipal and domestic water utilization will experience efficiency gains and decline further over the next

20 years, from 14% today to 12% in 2030 (Roland Berger 2011). The educational initiative, "Global Water Forum," calls for immediate action. Even though a number of initiatives have already taken place, they favor a more strategic approach that can save cost and accelerate favorable outcomes.

On our planet, there is also a finite amount of land to be used for agriculture, production facilities, housing, and mining and to maintain the balance of humankind, wildlife, and vegetation. In accordance with current demographic trends and global population growth, an increase of 60% in demand for food, feed, and fiber is projected by 2050. The expansion in the agricultural area is limited due to the lack of suitability for agriculture. The ecological, social, and economic costs of refitting ground would be very high. The Food and Agriculture Organization (FAO) of the United Nations states that "sustainable management of the world's agricultural soils and sustainable production have therefore become imperative for reversing the trend of soil degradation and ensuring current and future global food security" (FAO 2015).

Efforts for more sustainability are notable along the supply chain: the transport sector is implementing energy efficient technology, ranging from low-emission engines to electric-powered vehicles and telemetric concepts to reduce emissions. In 2015, the *London Underground* tested an inverter system that converts braking energy into power. The test lasted 5 days, resulting in capturing a full megawatt hour – enough to power more than 100 homes. The experiment was part of an initiative that considers ways of seeking low energy consumption options for the expanding Tube. In addition, it would save the London Underground money: if applied to the entire Tube system, the system would result in savings of about $9 million. The eco-friendly by-product is that the regenerative braking doesn't produce the heat of conventional friction braking; the tunnels themselves stay cooler, requiring less energy expenditure on climate control (Gibson 2015).

Truck platooning and self-driving vehicles will be major breakthroughs for energy efficiency and the reduction of carbon emissions. Transport network design and optimization also plays an important role. Examples include the utilization of a broad range of delivery options, such as lockers, drop boxes, or stores, to avoid empty runs on the last mile and failed deliveries. Freight exchange platforms, where carriers can find freight/back load or post their available transport capacity on the system to reduce empty vehicle mileage, will contribute too. Governments support the efforts. In 2009, the Federal Highway Administration, part of the U.S Department of Transportation, launched the Empty Miles Program – Matching Empty Containers with Available Freight; Canada established a similar program the preceding year. The programs proved to be successful: by taking advantage of the service, American retailer *Macy's* eliminated 11% of their empty miles year-on-year in 2010, while experiencing an increase of 30 backhaul loads per week or a projected 1500 loads per year. In its first year of existence, the participating US industry reduced greenhouse gas emissions by over 200 tons and increased backhaul revenue by 25% (Smith 2010).

Packaging is increasing, on the other hand, particularly in the fast-growing area of home deliveries. Reducing and reusing packaging, including developing standards regarding the environmental impact of packaging and products, is becoming an urgent need. The lighter the packaging, the less energy is needed for transport. The Sustainable Packaging Consortium, engaged in finding sustainable solutions, aims to improve the sustainability of packaging across the different value chain stages and focuses on driving improvement around the three Rs (redefine, reduce, reuse) (The Sustainability Consortium 2016).

The marine transport sector is taking on its share of responsibility. The IMO's Marine Environment Protection Committee has introduced emission control areas (ECA) in the Baltic Sea, and the North Sea in the USA and Canada and the US Caribbean. Some shipping companies have decided to install "scrubbers" to remove some particulates and/or gases from industrial exhaust streams. Instead of using bunker liquids, marine vessels turn to LNG (liquefied natural gas). Depending on the type of engine used, air pollutant emission reduction is significant, including nitric oxygen (NO) reduction of 85–90%; the use of LNG could result in a 20% decline in greenhouse gas emissions (World Port Climate Initiative 2016). The USA might find itself emerging as a beneficiary of this development. Even though gas prices are fairly low at the moment, experts predict the market to reach its lowest point in 2018 and then tighten up again. The liquid gas industry expects the USA to be one of the biggest three suppliers of LNG by 2020 (Domm 2016).

Clean Cargo Working Group (CCWG) is a global carrier-shipper initiative dedicated to improve the environmental performance of ocean cargo. More than 80% of global container shipping capacity is represented by the group; most of the top 30 carriers are members. According to a statement published in 2015, the average CO_2 emissions in global container shipping trade declined 8.4% in 2014 year-on-year. Judging from the baseline in 2009, the emissions fell 29%, based on a tool that measures, evaluates, and reports CO_2 impact of global goods transport. Members can compare different routes and choose the cleaner one (Burnson 2015).

Slow steaming – that is, operating ships at lower speed – reduces fuel and carbon emissions. *Maersk Line* introduced the concept in 2007. According to the leading container shipping line, slow steaming is contacted at 18 knots; lower speeds are called super slow steaming. The impact on the environment is notable: lowering engine speed by 10% cuts engine power by 27% and reduces the overall energy needed for the voyage by 19%. Slow steaming vessels are more likely to arrive at port on schedule too and are thus more reliable.

Aviation contributes around 2.5% of all emissions. In February 2016, the International Civil Aviation Organization (ICAO), a UN agency dedicated to ensuring safe and orderly growth in international air transport, agreed on international standards to limit carbon emissions from aviation in an ongoing effort to limit gases that trap heat in the Earth's atmosphere. Participants agreed on reducing their fuel consumption by an average of 4% compared to 2015 levels by 2028 (Maza 2016).

A tendency towards modular split, or shifting an increasing amount of cargo onto train transport, could also help with operating more fuel efficiently. The share of EU-28 inland freight that was transported by rail was 18.2% in 2015; in the USA, the percentage of freight shipped by rail accounted for 10% of the freight transport sector (road: 60%). But the amount shipped by rail, when measured by weight, is 38%. Compared to tractor-trailers, trains have become more effective in fuel efficiency.

How Resource Scarcity and Energy Shifts Impact Global Supply Chains

There are numerous initiatives that create a more sustainable and resource-efficient supply chain, such as eco-friendly vehicles, warehouses, or office buildings, which lower emissions of CO_2. Slow steaming is important, although adopted in face of rapidly rising fuel oil costs. Aviation and truck/tractor-trailer manufacturers are designing engines that are increasingly energy efficient and sustainable. Auto and commercial vehicle manufacturers are testing platooning and self-driving vehicles. A car that runs on compressed air is in development. The *AIRPod* could be the solution to reduce pollution, its creator *MDI*, based in Luxembourg, claims. The production of the vehicle in the USA is supposed to take place in Hawaii (Zero Pollution Motors 2016).

Transportation and delivery companies implement measures and use digital platforms to reduce empty mileage. Shippers and supply chain service providers are introducing joint programs to reduce energy needs and emissions. In the future, we are likely to see fuel-efficient and sustainable modes of transportation increase even more.

Initiatives from supply chain players also include reducing the amount of packaging or its weight. This will save energy and cause less emissions, while reducing costs due to the lower weight transported. Design plays a vital part as innovative, lighter materials can be applied to packaging. This has an impact on the storage of packaging and goods as well as on the repurposing of the packaging.

3.1.12 Climate Change

Nearly 25% of all greenhouse gas (GHG) emissions globally are transport-related. According to the Inventory of U.S. Greenhouse Gas Emissions and Sinks 1990–2011, the transportation end-use sector, including cars, trucks, commercial aircraft, and railroads, was responsible for 27% of GHG emissions in 2011. Within this sector, light-duty vehicles (including passenger cars and light-duty trucks) produced 61% of GHG emissions, while medium- and heavy-duty trucks contributed 22% of emissions (United States Environmental Protection Agency 2013).

In the EU, road transport contributed about one-fifth of the total emissions of CO_2 in 2015. Light-duty vehicles, cars and vans, produce around 15% of the EU's emissions of CO_2; heavy-duty vehicles, trucks and buses, are responsible for about one-quarter of CO_2 emissions from road transport in the EU and for some 6% of total EU emissions (European Commission 2016a, b). The IEA forecasts that trucking activity will double and air travel could increase fourfold by 2050 (International Energy Agency 2009).

Not just carbon dioxide is released into the air. Nitric oxide (N_2O), which is responsible for asthma and other respiratory diseases, is emitted from agriculture, transportation, and industry activities. The gas also has a much larger impact on global warming: the impact of one pound of N_2O is almost 300 times that of one pound of carbon dioxide. Globally, N_2O contributes 6% to overall emission of greenhouse gases. In the USA, N_2O emissions have increased by about 8% between 1990 and 2013 (United States Environmental Protection Agency 2016a, b, c).

Airborne nitric oxide could be reduced by an increasing introduction of pollution control technologies, such as catalytic converters to reduce exhaust pollutants from passenger cars. Also, utilization of other gases such as F-gas, which has a global warming potential (GWP) of 1 in contrast to N_2O, which has a GWP of 298, could be facilitated. Among the answers for reducing CO_2 emissions are purchase of energy and water efficient appliances, recycling and reuse, and teleconferencing to name just a few.

Ways to reduce CO_2 emission are not new in the transport industry. Many organizations have established regulations. The Environmental Protection Agency (EPA) and the National Highway Traffic Safety Administration (NHTSA) in the USA want to promote the production of a new generation of clean vehicles, through improved fuel use from on-road vehicles and engines. The agencies have adopted GHG regulations for heavy-duty engines and vehicles, which are yet to be lawfully passed (United States Environmental Protection Agency 2016a, b, c).

The EU is considering applying a CO_2 emission limit to heavy-duty vehicles too. Emissions from new cars and vans have been successfully reduced under EU legislation. Studies suggest that state-of-the-art technologies can achieve cost-effective reductions of at least 30% in CO_2 emissions from new heavy-duty vehicles (European Commission 2016a, b).

While CO_2 emissions in developed countries are expected to decline by 14%, CO_2 emissions in emerging countries are likely to increase by 38% to a global share of 68%. Already, some Asian countries are becoming concerned by this development. China, the world's largest GHG emitter, approved a development plan for climate change. Among the measurements is the objective to speed the rollout of more fuel-efficient cars and new energy sources. China has also set sector-specific energy efficiency standards. For instance, new commercial buildings must comply with building codes on energy use. Closing inefficient factories and power plants is also on the agenda (Center for Climate and Energy Solutions 2016). The US and Chinese governments have forged a partnership on energy and climate change and in 2014 jointly announced measurements to reduce net GHG, which include the launch of a climate-smart/low-carbon cities initiative and promoting trade in green goods (The White House 2014).

Water is another resource that must be used efficiently and in an environmentally safe way. The Global Water Forum points out that access to an improved water source does not always mean access to safe water. Micro-pollutants, such as medicines, cosmetics, cleaning agents, and biocide residues, are growing concerns (Global Water Forum 2012). By 2025, there might be more plastic than fish in the oceans. According to an Ellen MacArthur Foundation report, plastic production has increased 20-fold since 1964, reaching 311 million tons in 2014. The output is expected to double again in the next 20 years and almost quadruple by 2050. Despite the growing demand, just 5% of plastics are recycled effectively, while 40% end up in landfills, and one-third in fragile ecosystems such as the world's oceans (Ellen MacArthur Foundation 2016). Steps have been taken to reduce the distribution of plastic, for example, in a form of plastic bags.

Impacts of ecologically harmful production processes and increasing climate change are becoming apparent all across the world: dead lakes, droughts, floods, and sea-level rise. January of 2016 was the hottest January globally since the measuring of temperatures started in 1880 (Holthaus 2016). Extreme weather conditions, such as heavy rain or above-average temperatures, cause problems: storms cause delays in airfreight; extreme heat waves affect railways and cause pavement to soften and expand, making it more rutted and prone to damage; flooding causes delays in both road transport and airfreight. In the USA, approximately 60,000 miles of coastal roads are already exposed to flooding from coastal storms and high waves. According to the EPA, some of the busiest airports in the USA are located in low-lying coastal areas (United States Environmental Protection Agency 2016a, b, c). In some areas, the land is less moist due to drier winters, which causes wildfires that start earlier in the year and last longer. The 10.1 million acres that burned in the USA in 2015 were the most on record (Richtel and Santos 2016).

It is foreseeable that costs for maintaining transportation routes will increase in the future. In 2008, the National Research Council's Transportation Research Board conducted an in-depth study of the effects of climate change on land, marine, and air transportation in the USA. The board advised policymakers to incorporate climate change into investment decisions and improve communication (Transportation Research Board 2008). In the emerging and developing countries, intact infrastructure is vital to economic growth and wealth. Governments have to adapt their fiscal planning to not only building but also maintaining infrastructure.

Time is of the essence and might be running out: the iconic "Doomsday Clock" (Gannon 2015) has been set at 3 min to midnight. Governments and citizens are gradually realizing how global threats can destroy local lives. Climate change is probably the most prominent evidence that a collective effort is required to protect our lives and the lives of future generations.

How Climate Change Impacts Global Supply Chains
One of the most relevant value propositions to the well-being of future generations of players upstream and downstream in the supply chain is to protect the environment, particularly the climate, as much as we can. Supply chain design has to be assessed for its environmental impact. Chief supply chain officers have to acquire the necessary knowledge and tools of collaboration to work towards climate-neutral solutions. Solutions have to be both cost-efficient and environmentally sound. Collaboration and innovation across and beyond the entire supply chain ecosystem are required.

Shocks to the global supply and value chain can be severe. Lives, countries, cities, infrastructure, economic wealth, and prosperity are at risk – through flooding, erosions, or other natural disasters. Extreme weather disrupts all modes of transport. Natural disasters destroy significant value along the supply chain and have already put many out of business. Experts expect that the future will be even more violent.

3.1.13 Global Risks

Limited fertile land, scarce safe water, pollution, geopolitical tensions, social unrest, military conflicts, terrorism, natural disaster, pandemics, and cybercrime challenges are growing, and the aggregate of risks raises concerns. According to the "Global Risk Report 2015," provided by the World Economic Forum, the respondents of the global survey stated that their greatest concern is the flair-up of interstate conflicts (90%). With respect to the supply chain, nearly one-third (29%) of the executives asked saw the most notable impact in currency fluctuation and geopolitical risks. *Time* magazine listed as top geopolitical risks the politics in Europe and Russia, the effect of the China slowdown, the weaponization of finance, as well as ISIS, beyond Iraq and Syria (Bremmer 2015).

The Uppsala Conflict Data Program, at Uppsala University in Sweden, currently lists 120 armed conflicts, 414 non-state conflicts, and 206 one-sided violent incidents in 2014. Non-state conflicts, referring to the use of armed force between two organized armed groups, have particularly increased in the Middle East and in Africa in the recent years but is also picking up in Europe and the Americas (Uppsala University 2016).

Terrorist attacks aim at the lifelines of our economies. For now, such attacks have been limited: according to Daniel Ekwall's review of the official terrorist statistics from MIPT Terrorism Knowledge Base, transport activities represent only 4% of the targets in 2006 and 5% in 2007 (Ekwall 2012), as low a number as one can find. But the number of supply chain-related attacks has increased steadily over the past decade, reaching 3299 attacks in 2010, PwC found.

When terrorists do attack supply chains, the consequences could be disastrous. Seventy to eighty percent of the world's oil flows through only three major checkpoints: the Suez Canal, the Strait of Hormuz, and the Straits of Malacca (see graph) (Harvard Business School 2009). The closing of one or more of these passages would significantly disrupt the global energy supply chain. The Strait of Hormuz is a particular cause for concern: situated between the Arabian Peninsula and Iran, almost 20% of global oil trade passes through there (The Encyclopedia of Earth 2016).

Major freight hubs might be targets for terrorist attacks too. About 14.8% of containerized and airfreight traffic moves through the Hong Kong-Shenzhen freight cluster (PricewaterhouseCooper 2011). The Port of Shenzhen is one of the most important ports for China's international trade and therefore key to the supply of many factories and customers in the world.

Highway computer networks and traffic management systems are vulnerable to terrorist attack and could be hacked and disrupted to create chaos. This could make the transport on national road systems and in cities impossible. Planes could collide. Each time aircraft enter an airport, planes are at risk because of the sharing of information across a number of electronic systems (Collins, 2013). Tracks in rail traffic could also be reset to cause trains to crash.

We need to take adequate precautions to protect the flow of goods. Some governments already understand the threat. In 2009, President Obama declared digital infrastructure a strategic national asset (U.S. Department of Homeland Security 2013). The government of Canada has formulated a counter-terrorism strategy stressing the need for an integrated approach throughout all levels of government, law enforcement agencies, the private sector, and citizens, in collaboration with international partners and key allies, such as the USA (Government of Canada 2013). It is to be expected that after the Paris, Brussels, Nice, and Munich events, governments have tightened measures. Beyond a certain point, though, this can bring significant negative consequences for the global supply chain.

The private sector should follow suit, and the first necessary step is to accept the likelihood of such attacks. That is not the case often enough. A report prepared by the Supply Chain Management Faculty at the University of Tennessee found that while two-thirds of supply chain companies employed risk managers, virtually all of their internal functions ignored supply chain risk. In light of the current global developments, that is a risky position to take.

Terrorism is not limited to certain countries or religious groups. In addition, it cannot be placed into the hands of organized terrorist groups alone. Individuals who feel the urge to act on their behalf can become terrorists at any time. The possibility of having to fend off biological, chemical, or other kinds of aggression is increasing; furthermore, terrorists take a growing interest in global communication systems and infrastructural networks, both in regard to energy and transport, because the impact on global economy would be more harmful than attacks on home soil.

In 2011, when the Arab Spring spread like wildfire across several Arabic countries, many companies were reluctant to take action. How should they act in light of demonstrations run down by military forces, shutdown of the Internet and cellular networks, and collapse of public transportation, all putting employee's lives at stake? Some manufacturers stopped production on short notice and withdrew from the affected countries. Other companies stayed and tried to maintain the supply chain.

One example is automotive supplier *Leoni*, which produces cables, wires, and wiring systems and which is a provider of related development services for international clients in Egypt. In the region, Leoni runs three production plants and employs approximately 12,000 people. When the Arab Spring reached Egypt and caused public transport to collapse, workers were not able to get to work, or they stayed at home in order to protect their belongings from plunder. In addition, the three-shift system collided with curfew. In order to keep production going, Leoni had to act fast: shifts were adapted for curfews, a bus shuttle system was established that safely took employees to the production plant and back home, and food and beverages were provided. Communication was set up via text messages and radio, which announced the bus schedule. The supplier's aim: as long as the employee's safety was guaranteed, Leoni would not withdraw from the region and would keep up production. Leoni was able to ensure production in their two Egypt sites, even though it had to face severe impairments, when the region was drawn into the Arab Spring movement. They had also increased production in excess of demand prior to the riots in Egypt. Therefore, they were able to meet customer demand with the buffer, which was substantial. The supply chain also had to be adapted. Because marine transport had been restricted and temporarily terminated in the Port of Cairo, urgently needed parts had to be shipped in by airfreight (Leoni 2011).

Most businesses are aware of the fact that ICT is a vital part of their asset and reliable supply chain. Today, the majority of information from orders and capacity to customer data is shared digitally across the ecosystems with many stakeholders. Information is a production factor and source of competitive advantage. Hence, information is a target of economic and other crimes. According to Verizon's 2015 Data Breach Investigations Report, there were 2122 confirmed data breaches in the previous year at organizations in 61 countries (Verizon 2015). PricewaterhouseCoopers (PwC) reports in their study "2015 Information Security Breach Survey" that both large and small organizations in the UK had experienced an increase in the number of security breaches[11]: 90% of large organizations reported that they had suffered a security breach, up from 81% in 2014; security breaches at small organizations went up from 60% in 2014 to 74% in 2015.

In the first quarter of 2015, McAfee Labs saw a 165% increase from the previous quarter in new ransomware. According to the FBI, criminals are netting an estimated $150 million a year through rogueware, ransomware, and fake antivirus scareware scams. The hard-to-detect CryptoWall, which has been responsible for 992 ransomware attacks reported to the agency since it appeared in April 2014, continues to be a threat at the moment. The FBI says CryptoWall ransomware attacks alone have cost US business $18 million in the past year, not including indirect costs and unreported attacks (Ashford 2015).

The CERT-UK, the UK National Computer Emergency Response Team, formed in March 2014 in response to the National Cybersecurity Strategy, lists four main cybersecurity risks within the supply chain: through website builders, third-party software providers, third-party data stores, and so-called watering hole attacks, which work by identifying a website that is frequented by users within a targeted organization or even an entire sector and compromising that website to enable the distribution of malware (CERT-UK 2015). Mobile apps also pose a "growing and distinctive risk" to companies, a security study by Hewlett Packard Enterprise's "Cyber Risk Report 2016" found. Approximately 75% of the scrutinized mobile applications exhibited at least one critical or high-level security vulnerability, compared to 35% of the nonmobile applications (Weldon 2016).

---

[11]Security breach is any incident that results in unauthorized access of data, applications, services, data, and other networks and/or devices and assets.

From a biological perspective, the introduction of invasive foreign species to a society also poses a risk. One example is the Zika virus, which started to spread in early 2016. First isolated in 1947 in the Zika Forest of Uganda, the virus began to spread eastward across the Pacific Ocean to French Polynesia, then to Easter Island in 2014. In 2015, it reached Mexico, Central America, the Caribbean, and South America, where the Zika outbreak reached pandemic levels in 2016. The virus affects pregnant women and causes birth defects, miscarriages, and microcephaly. Other possible symptoms and effects are still being researched. The Ebola virus has been even deadlier. First discovered in 1976 in what is now the Democratic Republic of Congo, Ebola stretched out across Western Africa in 2013. Worldwide, there have been 28,639 cases of Ebola virus disease and 11,316 deaths as of 2015 (World Health Organization 2016).

But of course, humans are at least as much of a risk to wildlife as they are to us. Many species of wild animals and plants, such as elephants, tigers, rosewoods, and redwoods, are being driven to extinction by poaching. Between 2007 and 2014, rhino poaching in South Africa increased an incredible 9000%. Iguanas, turtles, and marmosets, among others, are caught for the illegal pet trade. Whole forests of rosewood trees have been illegally felled in Brazil, Cambodia, Benin, and Madagascar to feed the demand for luxury furniture in China. The illegal wildlife trade has a value between $7 billion and $23 billion annually.

The transportation and logistics sector has a critical role to play in identifying illegally poached material and arresting its movement along the supply chain. In March 2015, representatives from 17 companies, which account for 95% of the Chinese courier market, including EMS, DHL, FedEx, and SF Express, pledged zero tolerance for illegal wildlife trade. Many airlines have banned hunting trophies completely – including Air Canada, Air France, British Airways, Etihad, Lufthansa, Qantas, Singapore Airlines, and Virgin Atlantic – while others have banned trophies of protected animals including Delta, United, and American Airlines (Lehmacher 2016).

Eventually, it would be better to inform and empower park rangers and communities to stop poaching. Technology such and satellites and drones can allow for constant monitoring, stopping the crime at the root. Until then, blocking the route to market can reduce the value of such ill-gotten prizes.

Mitigating the risk of cyber or terrorist attacks, pandemics, and wildlife crime requires not only supply chain visibility but also staying on top of the information and developments within and around the ecosystem. One way to do this is to monitor social media networks closely. People often communicate worries and events with friends rather than with official authorities. Yet we will never live in a 100% secure world. The aim is to minimize risks and the potential consequences. IT solutions, such as platforms for the exchange of data and collaboration across the supply chain ecosystem, real-time monitoring of supply chain activities, screening of employees and partners, analysis of big data, and the use of predictive analytics, seem to be necessary and useful measures in today's volatile world.

How Global Risks Impact Global Supply Chains

Cyberattacks, terrorism, pandemics, and many other risks pose a significant threat to the supply chain ecosystem and consequently to the global economy and our lives. But the IoT will help to reduce global risks, and so will blockchain. Visibility is a competitive edge and differentiator. With high visibility and traceability of actions, it will be much harder to execute attacks or environmental offenses. Piracy and corruption will be more difficult too.

Big data, social media, and predictive analytics will help to identify trends and potential disruptions at the earliest stage or even before they emerged. In job interviews, moral values will become increasingly important to exclude employees with extremist beliefs. Since convictions might change over time, personal contact and a good social fabric is vital for security.

Business needs to improve supply chain performance and security at the same time. Mitigation of risks requires technology as well as awareness and training across the organizations and ecosystems. Contingency plans and risk management networks and procedures need to be established and kept up-to-date. Governments and international organizations need to support the effort. Collaboration between public and private sector, and also with citizens, is vital to fend off risks in today's more complex and interconnected world.

3.2   The World as It Is Changing

We feel the impact of today's volatile and rapidly changing world. Businesses, governments, and citizens are facing new threats and opportunities: digitization, robots, AI, social media, global warming, natural disasters, urbanization, population growth, geopolitical tensions, refugees, globalization X.0, etc. Change can increase fear and lead to despair or lift hopes and free creative energy. While the East largely embraces the dynamics, Western citizens are more skeptical. Western democracies and Western values are put to the test. The world faces increasing social tensions and global disintegration tendencies, which might in the end reverse many of the economic gains the global population has enjoyed.

Combining different economic and technological trends, the supply chain is developing in two opposite directions: it becomes both longer and shorter at the same time. While fast products and new technologies such as 3D printing lead to a shortening of the supply chain, increasing efforts of developing countries to participate in the global value network and international e-commerce ask for a longer supply chain. The supply and value chain transforms into a multilayered universe of platforms on local, regional, and global levels with a broad mix of direct stakeholders, contributing to different value chains at the same or different times.

Megatrends pose a continuous challenge to the supply chain ecosystem. Chief supply chain officers have to stay on top of the game: assess developments, weigh trends, make operational decisions, and contribute to the strategic debate on board level. Protectionism and trade barriers on the one hand, and new trade agreements on the other, influence business models and investment decisions and require new strategies. CSCOs bring valuable insight to the board room.

The supply chain has never been as complex as it is today. With complexity, the level of risk grows. Up to now, businesses were held accountable for their part of the supply chain; in the future, however, we will see a growing demand from governments and consumers for taking on responsibility for the entire supply chain, nationally and globally. CSCOs will need to act responsibly and protect the business from reputation and financial damage.

The increasing complexity will also affect local service suppliers. They will also need to assume responsibility for the entire chain like the global players. For example, even small businesses have to make sure that their business partners do not appear on either EU or US sanction lists, that their business complies with customs regulations, and that they only work with trustworthy partners and employees. Businesses need to check the sanction lists before inviting applicants for job interviews.

All business activities have to comply with regulations and commitments, from product design to production and packaging; to shipping, warehousing, and distribution; and to maintenance, repair, and repurposing. Standards have to be kept. Companies must respect and comply with binding frameworks based on national legislation, from regulations on foreign trade to self-imposed commitments, such as the Global Reporting Initiative (GRI) or the Global Compact, a set of universal principles established by the United Nations in order to create the world's largest sustainability initiative (UN Global Compact 2016). In the future, we will see growing standardization and legislation for nonfinancial reports. Sweden was among the first nations to introduce the legal requirement for state-owned companies to publish sustainability reports according to the GRI framework in 2007. Three years later, South African companies that listed on the Johannesburg Stock Exchange had to publish integrated reports that combined financial and nonfinancial information. In 2014, the EU issued a directive on corporate disclosure of nonfinancial and diversity information by companies employing more than 500 employees. This directive will come into effect on January 1, 2017.

Today, consumers increasingly make the rules. The consumer base and the general public become the supervisory authority, acting upon specific incidents and information and affecting business, government, and international organizations. For example, retailers that rely on manufacturers without labor safety regulations could be banished by boycotts. Conversely, companies that prove to run business in an ethical and sustainable way, for instance, could be recommended. Modern means of communication and social media are the enabler. Information has become part of customer experience; media turns into a vital touchpoint. Positive customer scorings and testimonials are vital for success and survival. According to a study published by the Wall Street Journal in May 2008, producer's ethical behavior will influence purchase decisions directly. Consumers punish unethical behavior by not paying the asked price, demanding a discount of up to 29%, and in turn, reward ethical conduct by paying up to 18% more (Trudel and Cotte 2008). Ethical behavior is the result of responsible thinking and doing. This includes respect towards life, nature, and the planet.

Literature

A.T. Kearney. (2015). *The 2015 global retail E-Commerce Index*. Available at https://www. atkearney.de/documents/856314/5715046/Global+Retail+E-Commerce+2015_vf.pdf/499bf80f-fb76-4a52-8f86-8ff09bb80e9e

Airbus. (2016). Accessed February 14, 2016, from http://www.airbus.com/innovation/future-by-airbus/the-concept-plane/the-airbus-concept-cabin/future-technologies/

AMAS. (2016). *Improve highway and bridge infrastructure and reduce congestion through a strong federal program with adequate funding*. Accessed April 29, 2016, from http://www. promover.org/files/govtaffairs/policy_papers/pp_highway.pdf

Ashford, W. (2015, June 26). *Ransomware costs business at least $18m, says FBI*. ComputerWeekly.com. Accessed Februaruy 16, 2016, from http://www.computerweekly. com/news/4500248823/Ransomware-costs-business-at-least-18m-says-FBI

Bonkenburg, T. (2016, March). *Robotics in logistics*. St. Onge Company/DHL. Available at http:// www.dhl.com/content/dam/downloads/g0/about_us/logistics_insights/dhl_trendreport_robot ics.pdf

Boston Consulting Group. (2011, December). *Global aging: How companies can adapt to the new reality*. Boston, MA: Boston Consulting Group

Bremmer, I. (2015, January 5). These are the top 10 geopolitical risks of 2015. *Time Magazine*. Accessed October 17, 2016, from http://time.com/3652421/geopolitical-risks-2015-ian-bremmer-eurasia-group/

Brennan, M., & Golden, J. (2015, October 2). *UPS, FedEx and DHL bet big on health-care logistics*. CNBC. Accessed February 10, 2016, from http://www.cnbc.com/2015/10/02/ups-fedex-and-dhl-bet-big-on-health-care-logistics.html

Burnson, P. (2015, September 03). *Ocean Cargo supply chain becoming greener, says BSR*. Supply Chain Manager. Accessed March 10, 2016, from http://www.scmr.com/article/ocean_cargo_supply_chain_becoming_greener_says_Bsr

Cage, S. (2016, March 8). *Flown out by drone*. Maersk.com. Accessed April 29, 2016, from http:// www.maersk.com/en/hardware/2016/03/flown-out-by-drone

Center for Climate and Energy Solutions. (2016). *Climate change mitigation measurements in the People's Republic of China*. Available at http://www.c2es.org/docUploads/International percent20Brief percent20-percent20China.pdf

CERT-UK. (2015). *Cyber-security risks in the supply chain*. Available at https://www.cert.gov.uk/ wp-content/uploads/2015/02/Cyber-security-risks-in-the-supply-chain.pdf

Chumley, C. K. (2014, July 18). *FedEx faces fines of $1.6B for shipping illegally sold drugs*. The Washington Times. Accessed September 15, 2016, from http://www.washingtontimes.com/ news/2014/jul/18/fedex-faces-fines-16b-shipping-illegally-sold-drug/

Collins, N. (2013, December 27). *Cyber terrorism is 'biggest threat to aircraft'*. The Telegraph. Accessed March 10, 2016, from http://www.telegraph.co.uk/finance/newsbysector/transport/ 10526620/Cyber-terrorism-is-biggest-threat-to-aircraft.html

Credit Suisse Research Institute. (2015, September). *The end of globalization or a more multipolar world?* (pp. 5–6).

Deloitte University Press, Business Trend Series. (2015). *Business ecosystems come of age* (p. 4)

DHL. (2015, January 26). *DHL successfully tests augmented reality application in warehouse*. Accessed September 15, 2016, from http://www.dhl.com/en/press/releases/releases_2015/ logistics/dhl_successfully_tests_augmented_reality_application_in_warehouse.html

Diehl, A. (2015, May 8). Opinion: Here is a three-pronged approach to pilot shortage. *Aviation Week*. Accessed September 15, 2016, from http://aviationweek.com/commercial-aviation/opin ion-here-three-pronged-approach-pilot-shortage

Domm, P. (2016, February 25). *U.S. exports of LNG mark a turning point in the market*. CNBC. Accessed March 08, 2016, from http://www.cnbc.com/2016/02/25/us-exports-of-liquified-nat ural-gas-mark-a-turning-point-in-the-energy-market.html

Dopart, K., & U.S. Department of Transportation. (2015, December 1). *U.S. DOT truck platooning research*. 2015 Florida Automated Vehicles Summit. http://www.dot.state.fl.us/planning/statis tics/fav/2015summit/Session8-Dopart.pdf

Ekwall, D. (2012). *Supply chain security—Threats and solutions* (p. 162). Intech. Available at http://cdn.intechopen.com/pdfs-wm/38967.pdf

Ellen MacArthur Foundation. (2012, August). *In-depth—Mobile Phones*. Accessed January 14, 2016, from http://www.ellenmacarthurfoundation.org/circular-economy/interactive-dia gram/in-depth-mobile-phones

Ellen Macarthur Foundation. (2015, January 23). *Project MainStream launches three new programmes*. Accessed Sepetember 15, 2016, from https://www.ellenmacarthurfoundation. org/news/project-mainstream-launches-three-new-programmes

Ellen MacArthur Foundation. (2016, January 19). *The new plastics economy: Rethinking the future of plastics*. Available at http://www.ellenmacarthurfoundation.org/assets/downloads/ EllenMacArthurFoundation_TheNewPlasticsEconomy_29-1-16.pdf

Escobar, P. (2015, February 22). *Tomgram: Pepe Escobar, inside China's 'new normal'*. TomDispatch.com. Accessed September 15, 2016, from http://www.tomdispatch.com/post/ 175959/tomgram%3A_pepe_escobar%2C_inside_china%27s_%22new_normal%22/

European Commission, Climate Action. (2016a). *Reducing CO2 emissions from heavy-duty vehicles*. Accessed February 15, 2016, from http://ec.europa.eu/clima/policies/transport/ vehicles/heavy/index_en.htm

European Commission, Climate Action. (2016b). *Road transport: Reducing CO2 emissions from vehicles*. Accessed February 15, 2016, from http://ec.europa.eu/clima/policies/transport/ vehicles/index_en.htm

Eurostat. (2016). *Passenger Cars in the EU*. Accessed February 12, 2016, from http://ec.europa. eu/eurostat/statistics-explained/index.php/Passenger_cars_in_the_EU

ExxonMobile. (2016). *The outlook for energy: A view to 2040*. Available at exxonmobile.com/ energyoutlook

FAO. (2015). *Soil is a non-renewable resource*. Rome. Available at http://www.fao.org/3/a-i4373e.pdf

Fisher, A. (2014, May 1). Wanted: 1.4 million new supply chain workers by 2018. *Fortune Magazine*. Accessed February 10, 2016, from http://fortune.com/2014/05/01/wanted-1-4-mil lion-new-supply-chain-workers-by-2018/

Flextronics. (2015, July 07). *Flextronics Unveils Flex Pulse™—A software-based, real time, comprehensive mobile view into supply chain events*. PR Newswire. Accessed September 15, 2016, from http://www.prnewswire.com/news-releases/flextronics-unveils-flex-pulse----a-software-based-real-time-comprehensive-mobile-view-into-supply-chain-events-300109705. html

Flood, C. (2015, June 14). *An extra $1tn a year for emerging market infrastructure*. Financial Times. Accessed February 08, 2016, from http://www.ft.com/cms/s/0/5bad10e0-100f-11e5-ad5a-00144feabdc0.html#axzz3zcBJ55mF

Frey, C. B. F., & Osborn, M. A. (2015, September). *The future of employment: How susceptible are jobs to computerization?* Available at http://www.oxfordmartin.ox.ac.uk/downloads/aca demic/The_Future_of_Employment.pdf

GameKlip. (2016). Last accessed December 11, 2015, from http://buy.thegameklip.com/

Gannon, M. (2015, January 22). Doomsday clock set at 3 minutes to midnight. Available at https://www.scientificamerican.com/article/doomsday-clock-set-at-3-minutes-to-midnight/

Gibson, D. K. (2015, October 2). *The train that powers its station*. BBC. Accessed March 10, 2016, from http://www.bbc.com/autos/story/20151002-the-train-that-powers-its-sta tion?ocid=fbindaut

Global Water Forum. (2012, May 21). *Water outlook to 2050: The OECD calls for early and strategic action*. Last accessed February 16, 2016, from http://www.globalwaterforum.org/2012/05/21/water-outlook-to-2050-the-oecd-calls-for-early-and-strategic-action/

Global Wind Energy Council. (2016). *Global status overview*. Accessed February 15, 2016, from http://www.gwec.net/global-figures/wind-energy-global-status/

Goldman Sachs. (2016). *The rise of China's new consumer class*. Accessed February 11, 2016, from http://www.goldmansachs.com/our-thinking/macroeconomic-insights/growth-of-china/chinese-consumer/

Gorenflo, N. (2016). *What if Uber was owned and governed by its drivers—The rise of platform cooperatives*. Economist. Accessed February 11, 2016, from http://evonomics.com/uber-shar ing-economys-biggest-threat/

Government of Canada, Public Safety Canada. (2013). *Building resilience against terrorism: Canada's counter-terrorism strategy*. ISBN:978-1-100-22422-0. Available at http://www.publicsafety.gc.ca/cnt/rsrcs/pblctns/rslnc-gnst-trrrsm/rslnc-gnst-trrrsm-eng.pdf

GT Nexus. (2016). *State of the global supply chain—A research report on the top issues facing industry executives for 2016 and beyond* (p. 6).

Hamburg Port Authority. (2016). Accessed February 12, 2016, from http://www.hamburg-port-authority.de/en/smartport/logistics/Seiten/Unterbereich.aspx

Harttmann, C. (2016, January 28). *Schweden: Digitales Lagerhaus à la Schenker*. Transport. Accessed February 10, 2016, from http://www.transport-online.de/Transport-News/Interna tional/15524/Schweden-Digitales-Lagerhaus-la-Schenker

Harvard Business School, The Centennial Global Business Summit. (2009). *Strategic responses to global terrorism* (p. 2). Available at http://www.hbs.edu/centennial/businesssummit/business-society/strategic-responses-to-global-terrorism.pdf

Holthaus, E. (2016, February 16). Earth kicks off 2016 with the most abnormally warm month ever measured. *Future Tense*. Accessed March 10, 2016, from http://www.slate.com/blogs/future_tense/2016/02/16/january_2016_was_earth_s_warmest_month_yet.html?wpsrc=sh_all_mob_tw_top

IBM. (2015, December 21). *How Wearable Tech is shaping logistics and the warehouse of the future*. Accessed February 14, 2016, from http://www.ibm.com/blogs/think/2015/12/21/ups-wearable-tech-iot/

IBM Global Business Services. (2010, October). *The smarter supply chain of the future* (p. 9 ff).

ICCT—The International Council on Clean Transportation. (2013). *European Vehicle Market Statistics, Pocketbook*. Available at http://www.theicct.org/sites/default/files/publications/EU_vehiclemarket_pocketbook_2013_Web.pdf

IEA. (2015). Available at https://www.iea.org/bookshop/708-Medium-Term_Renewable_Energy_Market_Report_2015

IFLN Network. (2016). Accessed February 08, 2016, from http://www.ifln.net/network.html

Inrix. (2015). Available at http://inrix.com/scorecard/

International Energy Agency. (2009). *Transport, energy and CO2—Moving towards sustainability*. Paris. ISBN:978-92-64-07316-6. Available at https://www.iea.org/publications/freepublications/publication/transport2009.pdf

International Energy Agency, World Energy Outlook. (2015). *Executive summary*. Available at https://www.iea.org/Textbase/npsum/WEO2015SUM.pdf

Ja-young, Y. (2015, December 30). *eBay becomes gateway for industrial exporters*. The Korea Times. Accessed September 15, 2016, from http://www.koreatimes.co.kr/www/news/biz/2015/12/488_194342.html

Kirkpatrick, N. (2015, March 31). *The world's most congested cities, by the numbers*. The Washington Post. Accessed February 11, 2016, from https://www.washingtonpost.com/news/morning-mix/wp/2015/03/31/the-worlds-most-congested-cities-by-the-numbers/

Lanks, B. (2015, April 27). *Your future office space might commute to you.* Bloomberg. Accessed February 12, 2016, from http://www.bloomberg.com/news/articles/2015-04-27/your-future-office-space-might-commute-to-you

Lavars, N. (2015, October 4). *Self-driving truck hits the highway in world first.* gizmag.com. Accessed February 12, 2016, from http://www.gizmag.com/daimlers-production-autonomous-truck-debuts-public-roads/39701/

Leflaive, X. (2012, May 21). *Water outlook to 2050: The OECD calls for early and strategic action.* Global Water Forum. Accessed February 16, 2016, from http://www.globalwaterforum.org/2012/05/21/water-outlook-to-2050-the-oecd-calls-for-early-and-strategic-action/

Lehmacher, W. (2015, June 12). *The next big thing in the sharing economy?* World Economic Forum Agenda. Accessed February 11, 2016, from http://www.weforum.org/agenda/2015/06/the-next-big-thing-in-the-sharing-economy

Lehmacher, W. (2016, September 28). *Wildlife crime: A $23 billion trade that's destroying our planet.* https://www.weforum.org/agenda/2016/09/fighting-illegal-wildlife-and-forest-trade/

Leoni. (2011, February 03). *Leoni largely normalised production in North Africa.* Accessed February 16, 2016, from https://www.leoni.com/en/press/releases/details/leoni-largely-normalised-production-in-north-africa/

Lockwood, R. (2016, April). *Truck platooning, past, present, and future.* Truckinginfo. Accessed April 30, 2016, from http://www.truckinginfo.com/article/story/2016/04/platooning-is-on-the-way.aspx

Lu, C. (2014, December 4). *Zara supply chain analysis—The secret behind Zara's retail success.* Trade Gecko. Accessed September 15, 2016, from https://www.tradegecko.com/blog/zara-supply-chain-its-secret-to-retail-success

Marchese, K., & Kelly, E. (2015, April 15). *Supply chains and value webs*, Business trends series. Deloitte University Press. Available at http://dupress.com/articles/supply-chains-to-value-webs-business-trends/

Markets and Markets. (2016). *UAV Drones market worth 5.59 billion USD by 2020.* Accessed April 29, 2016, from http://www.marketsandmarkets.com/PressReleases/commercial-drones.asp

Martin. (2015, February 25). *Mass customization: What, why, how, and examples.* Cleverism.com. Accessed September 15, 2016, from https://www.cleverism.com/mass-customization-what-why-how/

Maza, C. (2016, February 9). *Can the UN make air travel more sustainable?* The CS Monitor. Accessed February 15, 2016, from http://www.csmonitor.com/Environment/Energy/2016/0209/Can-the-UN-make-air-travel-more-sustainable

Mazzucato, M. (2015, December 15). *Who will fund the low-carbon revolution?* World Economic Forum. Accessed February 15, 2016, from http://www.weforum.org/agenda/2015/12/who-will-fund-the-low-carbon-revolution

Moore, J. F. (1996). *The death of competition: Leadership and strategy in the age of business ecosystems* (p. 26). New York, NY: HarperBusiness.

Motavalli, J. (2015, July 22). *Michigan has a fake city to test self-driving cars.* Mother Nature Network. Accessed February 12, 2016, from http://www.mnn.com/green-tech/transportation/blogs/michigan-fake-city-test-self-driving-cars

Murray, A. (2016, May 20). *GE's Immelt signals end to 7 decades of globalization.* Fortune. Accessed September 15, 2016, from http://fortune.com/2016/05/20/ge-immelt-globalization/

Myers, C. (2015, November 6). Is your company encouraging employees to share what they know? *Harvard Business Review.* Accessed February 10, 2016, from https://hbr.org/2015/11/is-your-company-encouraging-employees-to-share-what-they-know

Nash, K. S. (2016, July 14). *IBM pushes blockchain into the supply chain.* http://www.wsj.com/articles/ibm-pushes-blockchain-into-the-supply-chain-1468528824

PewResearch Center. (2013, May 29). *Breadwinner Moms.* Available at http://www.pewsocialtrends.org/files/2013/05/Breadwinner_moms_final.pdf

Pinelli, M. (2015, September 11). *Can entrepreneurship solve the youth unemployment crisis?* World Economic Forum. Accessed February 12, 2016, from http://www.weforum.org/agenda/2015/09/can-entrepreneurship-solve-the-youth-unemployment-crisis

PricewaterhouseCooper. (2011). *Transportation & Logistics 2030*, Vol. 4: *Securing the supply chain* (p. 7). Available at https://www.pwc.ch/user_content/editor/files/publ_trans/pwc_transportation_logistics_2030_vol4.pdf

Richtel, M., & Santos, F. (2016, April 12). *Wildfires, once confined to a season, burn earlier and longer.* New York Times. Accessed April 20, 2016, from http://www.nytimes.com/2016/04/13/science/wildfires-season-global-warming.html?_r=1

Rizzo, P. (2016, September 23). *World's largest mining company to use blockchain for supply chain.* http://www.coindesk.com/bhp-billiton-blockchain-mining-company-supply-chain/

Roland B. (2011). *The trend compendium 2030.* Munich. Available at http://www.rolandberger.com/media/pdf/Roland_Berger_Trend_Compendium_2030_Trend_2_Globalization_and_future_markets_20140501.pdf

Roland B. (2014). *Global Logistics Markets.* Munich, August 2014.

Ryan, A. S. (2012, June 29). *Top 10 air freight hubs.* Supply Chain – Digital. Accessed February 08, 2016, from http://www.supplychaindigital.com/logistics/2815/Top-10-Air-Freight-Hubs

Sachin R, S., & Saito, M. (2016, March 9). *Amazon to start air delivery network with leasing deal.* Reuters. Accessed March 10, 2016, from http://uk.reuters.com/article/us-air-transport-sr-amazon-com-idUKKCN0WB1LA

Shepardson, D., & Lienert, P. (2016, February 10). *In boost to self-driving cars, U.S. tells Google computers can qualify as drivers.* Reuters. Accessed February 15, 2016, from http://www.reuters.com/article/us-alphabet-autos-selfdriving-exclusive-idUSKCN0VJ00H

Smith, A. (2010). Innovative empty miles service—Enables efficient truck transport. *Logistics Quarterly Magazine, 16*(1).

Solar Energy Industries Association. (2016). *Solar industry data.* Accessed February 15, 2016, from http://www.seia.org/research-resources/solar-industry-data

Song, J. (2015, December 29) *Chinese look to Germany for milk powder.* Epoch Times. Accessed February 10, 2016, from http://www.theepochtimes.com/n3/1928978-chinese-look-to-germany-for-milk-powder/

Soper, S. (2016, February 9). *Amazon building global delivery business to take on Alibaba.* Bloomberg. Accessed February 11, 2016, from http://www.bloomberg.com/news/articles/2016-02-09/amazon-is-building-global-delivery-business-to-take-on-alibaba-ikfhpyes

Spirkl, K. (2016, January 07). *EU investiert Rekordsumme in Verkehrsprojekte.* Verkehrsrundschau. Accessed February 08, 2016, from http://www.verkehrsrundschau.de/eu-investiert-rekordsumme-in-verkehrsprojekte-1738307.html?fromSearch=true

Statista.com. (2016). *Number of vehicles registered in the United States from 1990 to 2013 (in 1,000s).* Last accessed February 12, 2016, from http://www.statista.com/statistics/183505/number-of-vehicles-in-the-united-states-since-1990/

Statistics Canada. (2016). *Population growth in Canada.* Accessed February 10, 2016, from http://www.statcan.gc.ca/pub/91-003-x/2007001/4129907-eng.htm

The Economist. (2015, October 31). *The great chain of being sure about things.* http://www.economist.com/news/briefing/21677228-technology-behind-bitcoin-lets-people-who-do-not-know-or-trust-each-other-build-dependable

The Economist. (2016, September 8). *Profits overboard.* Accessed September 15, 2016, from http://www.economist.com/news/business/21706556-shipping-business-crisis-industry-leader-not-exempt-profits-overboard?fsrc=scn/fb/te/pe/ed/profitsoverboard

The Encyclopedia of Earth. (2016). *Strait of Hormuz.* Accessed March 11, 2016, from http://www.eoearth.org/view/article/156265/

The Radicati Group Inc. (2015, March). *Email statistics report.* Available at http://www.radicati.com/wp/wp-content/uploads/2015/02/Email-Statistics-Report-2015-2019-Executive-Summary.pdf

The Sustainability Consortium. (2016). https://www.sustainabilityconsortium.org/packaging/

The White House, Office of the Press Secretary. (2014, November 11). *FACT SHEET: U.S.-China joint announcement on climate change and clean energy cooperation*. Accessed February 15, 2016, from https://www.whitehouse.gov/the-press-office/2014/11/11/fact-sheet-us-china-joint-announcement-climate-change-and-clean-energy-c

thepaypers.com. (2016). *Mobile commerce and online shopper behaviour in US*. Accessed February 11, 2016, from http://www.thepaypers.com/ecommerce-facts-and-figures/us/5

Tobe, F. (2015, November 2). *While Amazon doubles its number of warehouse robots to 30K, competing systems emerge*. Robhub. Accessed February 08, 2016, from http://robohub.org/while-amazon-doubles-its-number-of-warehouse-robots-to-30k-competing-systems-emerge/

Transportation Research Board. (2008). *Potential impacts of climate change on U.S. transportation*. Available at http://onlinepubs.trb.org/onlinepubs/sr/sr290.pdf

Trudel, R., & Cotte, J. (2008, May 12). Does being ethical pay? *The Wall Street Journal*. Accessed February 16, 2016, from http://www.wsj.com/articles/SB121018735490274425

U.S. Department of Homeland Security. (2013, February 13). *Fact sheet: Executive order on cybersecurity/presidential policy directive on critical infrastructure security and resilience*. Accessed March 11, 2016, from https://www.dhs.gov/news/2013/02/13/fact-sheet-executive-order-cybersecurity-presidential-policy-directive-critical

U.S. Energy Information Administration. (2016). Form EIA-886 "Annual survey of alternative fueled vehicles." Available at http://www.eia.gov/renewable/afv/

UN Global Compact. (2016). https://www.unglobalcompact.org

UNEP. (2016). *Cities and climate change*. Accessed February 11, 2016, from http://www.unep.org/resourceefficiency/Policy/ResourceEfficientCities/FocusAreas/CitiesandClimateChange/tabid/101665/Default.aspx

United Nations. (2015). *World population prospects, 2015 revision*.

United States Census. (2010, May). *The next four decades, The older population in the United States: 2010 to 2050, population estimates and projections*. Available at https://www.census.gov/prod/2010pubs/p25-1138.pdf

United States Department of Labor, Data & Statistics. (2016). *Women in the labor force*. Accessed February 10, 2016, from http://www.dol.gov/wb/stats/stats_data.htm

United States Department of Transportation, Intelligent Transportation Systems Joint Program Office. (2015). Accessed February 12, 2016, from http://www.its.dot.gov/landing/strategicplan2015.htm

United States Environmental Protection Agency. (2013, September). *U.S. transportation sector greenhouse gas emissions 1990-2011, EPA-420-F-13-033a* (p. 1)

United States Environmental Protection Agency. (2016a). *Climate change/impacts/transportation impacts*. Accessed February 15, 2016, from http://www3.epa.gov/climatechange/impacts/transportation.html

United States Environmental Protection Agency. (2016b). *Overview of greenhouse gases*. Accessed February 15, 2016, from http://www3.epa.gov/climatechange/ghgemissions/gases/n2o.html

United States Environmental Protection Agency. (2016c). *Transportation and air quality/transportation and climate/regulations and standards*. Accessed February 15, 2016, from http://www3.epa.gov/otaq/climate/regulations.htm

Uppsala University. (2016). Department of Peace and Conflict Research: http://www.pcr.uu.se/research/UCDP/

Valenzuela, J. (2016, January 20). *Arcade city: Decentralized, blockchain-based answer to Uber*. The Cointelegraph. Accessed February 14, 2016, from http://cointelegraph.com/news/arcade-city-decentralized-blockchain-based-answer-to-uber

van Marle, G. (2016, February 18). *Going underground—The next logistics revolution*. The Load Star. Accessed March 08, 2016, from http://theloadstar.co.uk/going-underground-the-next-logistics-revolution/

Verizon. (2015). Available at http://www.verizonenterprise.com/DBIR/2015/?utm_source=pr&utm_medium=pr&utm_campaign=dbir2015

VOLVO. (2015). Pressrelease 02/19/2015. Accessed February 12, 2016, from https://www.media. volvocars.com/global/en-gb/media/pressreleases/158276/volvo-cars-presents-a-unique-system-solution-for-integrating-self-driving-cars-into-real-traffic

Vorabutra, J.-A. (2016, October 3). *Why blockchain is a game changer for the supply chain.* http:// www.supplychain247.com/article/why_blockchain_is_a_game_changer_for_the_supply_chain? utm_content=bufferf7878&utm_medium=social&utm_source=twitter.com&utm_campaign= buffer

Wallis, K., & Zawadzki, S. (2014, October 9). *Maersk and MSC ship-sharing pact gets U.S. clearance.* Reuters. Accessed February 08, 2016, from http://www.reuters.com/article/ maersk-alliance-idUSL3N0S35A820141009

Walton, M. (2015, July 20). *UK government releases rules to get self-driving cars onto public roads.* ars technica UK. Accessed February 12, 2016, from http://arstechnica.co.uk/cars/2015/ 07/uk-government-releases-rules-to-get-self-driving-cars-onto-public-roads/

Weldon, D. (2016, February 29). *Data security threats growing, putting projects and innovation at risk.* Information Management. Accessed March 10, 2016, from http://www.information-management.com/news/security/data-security-threats-growing-putting-projects-and-innovation-at-risk-10028336-1.html

Whole Foods Market. (2010, August 16). *National survey shows organic foods now represent larger part of total food purchases.* Accessed March 07, 2016, from http://media. wholefoodsmarket.com/news/national-survey-shows-organic-foods-now-represent-larger-part-of-total-food

Wikipedia. (2016). *Motor vehicle.* Accessed September 15, 2016, from https://en.wikipedia.org/ wiki/Motor_vehicle

Wolfgang, N. (2016, February 10). *Google gets patent for delivering packages through autonomous vehicles.* newstechtoday.com. Accessed February 10, 2016, from http://www. technewstoday.com/28545-google-gets-patent-for-delivering-packages-through-autonomous-vehicles/

Woollaston, V. (2016, March 10). *'Self-driving ground drones' hit the London streets: Trials of the delivery robots begin in Greenwich before heading to New York.* Daily Mail. Accessed April 29, 2016, from http://www.dailymail.co.uk/sciencetech/article-3486188/Self-driving-ground-drones-hit-London-street-Trials-delivery-robots-begin-capital.html

Word Atlas. (2016). *The world's largest oil reserves by country.* Accessed February 14, 2016, from http://www.worldatlas.com/articles/the-world-s-largest-oil-reserves-by-country.html

World Health Organization. (2016, February 3). *Ebola situation report.* Accessed February 16, 2016, from http://apps.who.int/ebola/current-situation/ebola-situation-report-3-february-2016

World Port Climate Initiative. (2016). *Benefits of LNG.* Accessed February 15, 2016, from http:// www.lngbunkering.org/lng/environment/benefits-of-LNG

Zero Pollution Motors. (2016). *AIRPod.* Accessed April 20, 2016, from http:// zeropollutionmotors.us/

Circular Economy and Society

4

Since the beginning of time, the earth has operated as a regenerative system. Only in our most recent history has mankind come close to largely burning, burying, and poisoning the basis of our existence and that which future generations will need to live. Global industrialization endangers resources including forests, oceans, and the atmosphere.

We have not yet found a way to properly recycle our waste. Plastics end up in the oceans instead of being reused in subsequent supply and value chains. We are destroying the "lung" of our planet: rainforests once covered 14% of the earth's land surface. Today, they cover only 6%. Predictions suggest that the last rainforests could be consumed in <40 years. Indonesia has the world's highest rate of deforestation. From 2000 to 2012, the country lost more than 23,000 square miles of forest to logging, agriculture, and other uses. That is roughly the area of West Virginia (Arunarwati et al. 2014). The Indonesian government has tried to slow down the destruction, with only limited success so far.

Earth has the ability to heal itself, though. It has done so in its long history and will continue to do so. We can help in the process by limiting the strain we add to the environment. Besides, nature is a good teacher. People have copied Mother Nature's inventions since the earliest times. Fishermen designed nets based on spider webs. Economics and societal science have adopted numerous concepts from nature; mankind has copied many of nature's best ideas for its own good. As we are all natural beings, it is only logical that we elevate nature as a whole to be our role model.

Respect for people and the planet is at the core of regenerative thinking and acting. Respectful behavior towards colleagues, customers, and consumers; current, parting, or former business partners; suppliers, governments, and international organizations; and media and the broader society is generating goodwill towards organizations and individuals. This is key for achieving organizational missions, including attracting people, obtaining permits and licenses, investment approvals, and financing. Running a circular business which is conscious of the potential future

© Springer International Publishing AG 2017   113
W. Lehmacher, *The Global Supply Chain*, Management for Professionals,
DOI 10.1007/978-3-319-51115-3_4

effects of our doing should be a value and ambition of each responsible leader in today's world.

4.1   The Circular Concept

The circular economy (CE) fights the wasteful exploitation of resources. By applying circular thinking and principles, environmental pollution, violence, and poverty should become a thing of the past. The CE is a vision and should as such be embedded in the vision of leaders, companies, and other organizations as well as nations. Legal frameworks, regulations, private and public actions, initiatives, and programs should reflect the intent and goal to create and respect a regenerative system.

CE focuses on three key principles:

1. Extension of the initial use cycle by reprocessing and upgrades
2. Initialization of subsequent use period to resell product at secondary market (after making minor modifications and repairs)
3. Reuse of parts and products that cannot be resold or fixed/processed further by recovering and reuse of integrated materials (including parts and matter accrued while making upgrades and repairs or while processing)

Chief supply chain officers are of utmost importance in implementing the CE and have to play their role in the company's overall strategic thinking and decision-making. CSCOs drive the supply chain development within the given framework, contributing their share to reduce, reuse, remake, and recycle. Circular products and processes require circular design. Designers and decision-makers in business and government need to educate themselves on regenerative skills. They need to monitor trends, methodologies, and technological progress in order to be able to include innovative concepts and regenerative methods into new projects and upgrades of existing assets and products.

Challenges to the CE are manifold. In business there are problems such as protective thinking in new product development teams; quality of recovered materials; negative impact on brands when using recycled materials, e.g., in luxury products; pricing of materials and used parts and products; and simply the practical application of the circular concept, to name a few. In government, most concerns are those about safety and security, which come along with all new technologies, methods, products, and materials. Consumers also need to be convinced. In 2015, UK-based price comparison website uSwitch published the results of a consumer study conducted on British consumers' awareness and attitude towards telematics-based car insurance policies. More than half of the responders (55%) raised concerns. They feared being spied on or having their data sold to other companies (Salmon 2015).

This shows clearly that the CE can only be realized with the help of the entire society. With this in mind, we need to add the societal dimension to the concept.

Therefore, I will speak about the circular economy and society (CE&S). We must involve the consumers and citizens to understand and address their concerns regarding the practices and technologies at the core of the circular world.

4.2   The Rainmakers

Rainmakers are technologies and models which support the design and implementation of the CE&S. Digitization has brought many technologies which support circular practices. Platforms and IoT, for example, can be major drivers of a regenerative economic and societal system. The following are three main developments which are highly relevant to the CE&S.

4.2.1   Sharing Platforms

The modern economy happens largely on platforms. Platforms can track the use of assets and products and enable the integration of the different stakeholders involved. Platforms like *Amazon* and *eBay* are facilitating the reuse of goods by offering a secondary marketplace to private individuals and professionals. Sharing platforms improve matchmaking between supply and demand and the asset or product utilization. Hence, sharing platforms reduce the number of assets and products needed and produced.

Ridesharing, for example, reduces the need for owning a car and hence frees space and eases congestion. *Uber* and *Lyft* are mobility companies and pioneers in the world of ridesharing platforms. With UberRUSH, the fast, on-demand pickup and delivery couriers, and UberEATS, the food delivery service, Uber offers goods transport in addition to their passenger service.

Ridesharing services are only one example of the sharing economy. The sharing economy,[1] also called the on-demand economy or collaborative consumption, has driven many new business models aimed at better utilization of assets, such as planes and homes. On the many online marketplaces, private individuals and professionals can find the cheapest or most suitable flight, the most convenient cab, or the most comfortable lodging, instead of standard hotel rooms. Business customers and consumers suddenly got easy access to a universe of choice and can give and get scores and opinions on services and products.

The sharing economy has enjoyed remarkable growth. Some projections put the sector's revenues at $335 billion globally by 2025. The room-letting website *Airbnb* started in 2008. Seven years later, the service was available in more than 190 countries, and the company is valued at more than $20 billion. Uber, the ridesharing app, was founded in 2009. In 2015, it operated in more than 300 cities

---

[1]Sharing economy refers to the economic model in which customers are able to borrow or rent assets, goods, or service owned by someone else.

in over 60 countries. In the same year, it was valued at more than \$50 billion (Marchi and Parekh 2015).

The term "share economy," coined by Harvard-based economic theorist Martin L. Weitzman in 1984, was linked to value creation from the beginning. He postulates that economic wealth will increase with expanding collaboration. Originally, this did not refer to the peer-to-peer-based sharing of access to goods and services we see today, but was connected to the question of whether a fixed salary or performance-related compensation would lead to a higher degree of economic wealth. In 2009, the definition as we know it today was used at a next09 conference hosted by the Internet marketing agency SinnerSchrader. The development of the sharing economy is closely related to the growing popularity of the Internet; in terms of the megatrends, boundaries between business and private offers and use are blurring. Posting, retrieving, and exchanging data and information have become simple and fast since the emergence of apps.

The Internet has made the sharing economy scalable, but it did not invent it. The idea is not a recent phenomenon: in the agricultural sector, it is a long-established tradition to share farm machinery that was too expensive for one single farmer to afford. Farmers share with colleagues according to a fixed schedule. Today, farmers often offer their machinery along with their manpower and experience rather than having it sit in the shed; farm contracting is a way of turning assets/products into a service.

In the past, both individuals and businesses have used sharing, in the form of lending and renting, as a method of avoiding purchase of expensive assets and goods. Portable toilets for construction sites or events are common, as too are service tools and garden equipment rental. Co-working spaces can be used when needed. Parents can even rent toys for their kids, which reduces clutter and helps the environment because the toys will not be discarded as easily or become outdated.[2]

New services include the booking of drones. The Drone Operations Network (DON) by *Future Aerials* brings together drone operators and business customers who require aerial pictures for surveys or inspections of assets, ranging from infrastructure to mines (Future Aerials 2016). Today, there is virtually nothing that can't be rented and shared, artworks for home and office, furniture, cars, clothes, condos, and houses; streaming of movies and TV series is also a way to get around owning and increase using. Roofs can be rented for solar panel installation.

The sharing economy created many new business models and companies such as online rental marketplaces like *eRentz*, *GoRenty*, *Rentalic*, and *Rentongo*, to name a few. Data published by Deloitte University Press suggests that three out of four people believe they will increase their sharing of physical objects and spaces until 2020. In 2013 the *Economist* wrote that according to Rachel Botsman, the author of a book on the subject, the value of the consumer peer-to-peer rental market is estimated to be \$26 billion (The Economist 2013).

---

[2]For example, pley.com, toystruck.com

While we are currently wasting as much as 40% of our food, sharing of leftovers is a growing trend (Grunder 2012). In Germany, the nonprofit organization *foodsharing.de* strives to connect those who have edible goods to give away with those who want them. *LeftoverSwap*, a Seattle-based start-up, has developed an app that lets one offer leftovers to locals. Users sign up and get notified when there's free food to be had in their area. In May 2014, the service had some 10,000 users in U.S cities like New York and San Francisco, as well as in Europe, Australia, and Asia. California-based *CropMobster* offers a different spin on food sharing: the website allows farmers to post excess crop that would otherwise be sent to the compost; volunteers collect it for further distribution to charities (Braw 2014).

With these and many other approaches, the sharing economy makes more efficient use of resources and capacities. What is still voluntary today might soon become obligatory. Instead of trashing food past its expiration date, food retailers could be obliged to give away unsold produce under certain conditions. In France, for example, supermarkets larger than 4305 square feet (400 m^2) are already banned from throwing away food that can't be sold anymore. Instead, they have to give it to food banks and charities (Chrisafis 2016). In the USA, food wasting is also an issue, since supermarkets and grocery stores toss out $46.7 billion worth of food. Laws exist in the USA to encourage food donation, including the Bill Emerson Good Samaritan Food Donation Act, various tax deductions, and the US Federal Food Donation Act of 2008. In addition, charities and for-profit companies such as AmpleHarvest and CropMobster collect and distribute food. Collectively, these organizations provide food to 46.5 million people annually. Major food retailers in the UK are also offering ways to reduce waste: instead of buy-one-get-one-free promotions, Tesco has buy-one-get-one-later, and Sainsbury's is piloting a buy-one-give-one-free program.

Companies and citizens could offer capacities and unexploited resources. Various examples make clear that resourceful and responsible behavior is beneficial for the environment and for revenue, in the form of increased stakes and additional profit. In combination with corresponding infrastructure and funding programs, this could be sufficient incentive for initiating this development. Whether this would need to be obligatory remains to be discussed.

Naturally, the sharing economy is also important for logistics service providers. Amazon and start-ups like *PiggyBaggy* (Finland) and *Nimber* (Norway) are trying to establish themselves as delivery companies. The latter connect customers willing to pay travelers for transport and delivery of their parcels. Finland's PiggyBaggy is now looking to start business in Denmark and Germany. Nimber already has 27,000 registered users since starting in Britain from mid-2015 till early 2016 and is targeting about 100,000 by the year-end. In Asia, an app by *Lalamove Singapore* matches customers with van drivers in order to provide on-demand delivery service from many drivers, who are often self-employed and not overly busy.

The reward and disruptive potential is high. Experts predict that if Uber-daughter *UberRUSH* were to capture 10% of the UK's courier market, it would translate into a 700 million pound ($996 million) income. Reuters cites Credit Suisse analysts who estimate that, if *Amazon* delivered half of its own parcels in

3–5 years, it could mean an annual 3–5% revenue loss for DHL's parcel business (Prakash 2016).

We need a model that ensures fairness while preserving the sharing economy's economic and environmental benefits. Businesses that are based on sharing certain products or services are not under the same scrutiny as traditional businesses; hence, they often do not have the same obligations and costs compared to their counterparts. On the other hand, the sharing economy facilitates entrepreneurship and helps better asset utilization.

Another point to be resolved is that providers in the sharing economy are hard to grasp for tax authorities. The government will suffer tax revenue losses, which are then missing from the amount planned to spend on investments, which are also necessary to keep the sharing economy afloat. However, with growing digitalization, the traceability of any kind of business activity will be simplified and raised. The sharing economy has to be clearly defined. Competition law has to be adapted so as not to put regular businesses at a disadvantage. Bad regulation will lead to job cuts and a decrease in economic wealth.

4.2.2   Product-as-a-Service

Service is the future. Services allow for much better monitoring and control of assets and products throughout the entire use cycle. Thanks to IoT technology, the condition of the asset might be well documented, which would form a clearer basis for the decision to reuse, remake, or recycle.

In the future, parts of vehicles and machines or the entire assets might not be owned by users anymore, but by manufacturers or circular economy players drawing revenues and margin from reusing, leasing, maintenance, upgrading, remaking, and recycling. New ownership models and reuse can be observed in the auto industry already, in the field of starters and electric car batteries for automobiles. "A used battery and new battery perform very similarly on many performance metrics," said the head of *4R Energy*, the second-life battery joint venture between *Nissan* and *Sumitomo Corporation* in 2015 (Think Progress 2016).

Products can become services: product-as-a-service is a new category of business models, which better leverage the value-in-use, another label for product-as-a-service, through bringing along better ways of monitoring the use of assets and estimating residual value at any time. The service offers include sustainable maintenance, upgrade, and replacement options. Instead of the physical standalone product that may include after-sale services, customers buy time of use or number of times used or applications.

The potential is high. Today, after-sales services represent up to 50% of revenues for some leading manufacturers. By 2018, 40% of top 100 manufacturers of countable finished goods and discrete manufacturers and 20% of top 100 manufacturers of bulk products such as chemicals, food, and beverages and process manufacturers are estimated to provide product-service platforms (Ashton 2014). A product-service system (PSS) stands not only for the mixed offering of products and services, but describes a function-oriented business model providing sustainability of both consumption and production. Technologies such as the IoT are key enablers of the new business model category.

Sensor-equipped trucks register driving behavior and its impact on parts such as tires or brakes. Fleet managers are able to plan maintenance based on real usage, monitor driving habits to identify training needs, or analyze operating situations and patterns to optimize fleet performance. Some telematics on-board units indicate poor driver behavior in real time, asking the driver to adapt accordingly. This leads to higher cost efficiency and longer use of assets. At the end of the use cycle, the condition of the monitored assets or parts are known, and the value can be estimated based on the analysis of real and specific cycle use data.

Xerox, for example, leases its copiers and then remanufactures them for reuse all over the world, in line with the circular approach. Several concepts have been established in their product-service system (PSS). The first requirement is a *performance guarantee*: even though the product is located at the customer's site, the provider monitors its performance. Service agreements ensure a high uptime. This kind of product-as-a-service can be found when leasing a printer or renting server space for telematics solutions. *Life cycle costs* are the various costs that occur during the operating period of a machine, such as costs for maintenance and repair. They are usually included in this service. Some customers also choose this hassle-free, all-inclusive fee while leasing a car. The product-as-a-service model also allows *continued optimization*, the guarantee that customers have the latest version of an operating machine. Thanks to big data and cloud computing, data from sensor-based products is collected and used for remote diagnostics and fix services, creating a much more efficient and effective operation.

Software is no longer sold separately as an actual product but as a *continued update*, which has to be downloaded; the user pays for a period of time and the application within the software that he or she is using. In return, he or she will automatically receive the latest version of the program. Examples of software leased this way are Adobe Creative Cloud and Microsoft Office 365. Another service possibility is *pay-per-use*, wherein the customer does not buy or lease a product, but pays for it only when he or she is actually using it. Similar to software-as-a-service in IT, industrial equipment is sometimes offered on a pay-per-use basis. This service helps customers to reduce upfront capital expenditures, while affording the manufacturer a continuous service revenue stream. Often payment is calculated according to the number of manufactured goods.

Finally, some industries have developed systems of *chemical leasing (ChL)*. Instead of selling chemicals, the producer sells the functions performed by the chemical. Functional units (number of pieces, area coated, etc.) are used to quantify the payment. The main applications currently are in the cleaning, greasing/degreasing, cooling/heating, and paint industry.

Among the increasing number of product-as-a-service solutions, experts have identified innovative emerging applications, such as the industrial machine that replaces the need for an on-site administrator. Instead, technicians will be equipped with smartphone apps that keep them apprised of problems but free to deal with other tasks. Another service could be a replenishment service that delivers refills automatically based on machine status instead of requiring human supervision (Stackpole 2015).

Thanks to this trend, new modes of differentiation are arising. Services are more complex and not as easily replicable as products. The mixed business models will tap new market potentials and create additional sources of revenue and profit. Manufacturer will turn service provider and thus connect with a broader, but also more modern, customer base with high expectations concerning service and support. Given the global competition and availability of similar product everywhere, customers have become less loyal. Product-as-a-service is a new way to differentiate and engage with customers. Since maintenance costs can become a significant component in the value proposition, the design of products should consider the complete use cycle. Better and more cost-effective maintenance leads to a prolonged utilization and possibly to more and better remaking and recycling options.

4.2.3 The Circular Supply Chain

At the core of the CE is the circular supply chain (CSC). Without the CSC the CE&S would be unthinkable. It is the CSC which brings the goods to the use and defines the route towards repurposing: reuse, remake, and recycle.

The CE represents a regenerative system, where everything receives a subsequent purpose: all materials and goods, including those resulting from the production process, flow on and on. In the circular world, all products and assets are transformed without negative impact to people and planet, distributed without emissions, consumed responsibly, and handed over at the end of each use cycle for reusing, remaking, or recycling. At the end of one use cycle, materials and goods enter another supply and value chain or go back to the earth without doing any harm, immediately or in the future. One core component for putting this concept of continuous flow and transformation without harm into reality is the CSC, for it is the tool by which everything else links and moves.

The best way to understand the CSC is to take a look at the linear supply and value chain first. In short, the linear value chain operates on the principle take-make-waste. Consequences for resources, environment, and society are hardly taken into account at any stage – from procurement and production to use and disposal. For instance, in order to keep cost low, commodities are assembled without thinking about separation later. That is a key prerequisite for repurposing. The option for reuse is often already disregarded when the product is designed. The linear supply chain runs its course into wastefulness.

Seventy percent of all products and parts end up as waste. In 2013, Americans generated about 254 million tons of trash and recycled and composted only 87 million tons of this material, equivalent to a 34.3% recycling rate (United States Environmental Protection Agency 2016). Almost two-thirds of the resources that went into making products and parts were not conserved or repurposed in the supply and value chain. But businesses, federal/national governments, and cities have realized the importance of reducing waste and are looking for concepts and practical solutions.

Right here, the CSC comes in, aiming for waste prevention and subsequent reuse of products and continuous use of resources through remaking and recycling. Every step along the supply and value chain is designed meticulously – from materials selected to procurement, production, distribution, and reuse of out-of-use or malfunctioning products and parts. In its ideal state, waste and emissions are not generated in the production or in the use and reuse phases, during remanufacturing or recycling. Materials stay in a closed-loop cycle, and when finally given back to earth, they are nontoxic and do no harm. The zero waste, zero emission supply chain is the ultimate goal.

The CE also requires the support of the consumer. In the USA, 15% of consumers send back what they buy from e-commerce sites (merchants, 10%). Handling these returns triggers not only costs and time for retailers but also shows the way to innovative business models once more. *Optoro*, for instance, is a start-up that aims to make the process of sorting through unwanted or damaged gadgets, apparel, and other products more efficient. Optoro's business intelligence and data analytics software, based on a cloud platform, helps automate reverse logistics. Retailers find potential buyers for unwanted or excess items more quickly. Optoro also streamlines processes by eliminating the number of touchpoints, reducing the risk of potential damage (Clancy 2015).

This seems to be the way ahead. A survey by the Recycled Paperboard Alliance, who issues the "100% Recycled Paperboard" certificate, found that 61% of consumers are more inclined to purchase products from a company that uses recycled paperboard packaging.

Spotlight: Reverse Logistics

The *Reverse Logistics Magazine* defines reverse logistics as "the process of planning, implementing, and controlling the efficient, cost effective flow of raw materials, in-process inventory, finished goods and related information from the point of consumption to the point of origin for the purpose of recapturing value or proper disposal. More precisely, reverse logistics is the process of moving goods from their typical final destination for the purpose of capturing value, or proper disposal. Remanufacturing and refurbishing activities also may be included in the definition of reverse logistics" (Hawks 2006).

A study by the University of Nevada found that 40% of supply chain executives say that return management is not as important as other current issues; 34.3% of respondents say they do not have good reverse logistics systems in place. Since return rates range from 3 to 50% of total products shipped, depending on industry and business, this is an overlooked potential for competitive advantage. Reverse logistics offers the possibility to create advanced customer service (Cerasis 2015).

One example of effective recycling and use of waste comes from New Zealand. *PT Enviro Pallets* is a manufacturer of nestled pallets made entirely from recycled plastic waste. People can actually sell their plastic waste at the factory; by requiring at least 30 t of plastic waste a day, landfills are also spared from having to deal with plastic (Erviani Komang 2012). Importantly for export usage, the pallets are resistant to insects, fungi, and bacteria. The firm has expanded business to Indonesia.

Making money from waste has also proven to be successful in Alaska, where a fishing boat captain turns salmon skin into accessories. *Tidal Vision* launched a line of wallets in 2015. The firm is also working on an environmentally friendly way to extract a compound called chitin from crab shells to make chitosan (Wong 2015). This substance is used in agriculture and in medicine. Discarded fish guts, heads, tails, fins, skin, and crab shells all cause a problem to marine waters. Decomposing organic matter can drain oxygen nearby living species might require. In addition, this foul mass can bury other organisms or introduce disease and nonnative species to the local ecosystem.

Waste is considered value in the CSC. The CE offers room and opportunity for a multitude of innovative business models. Countries, companies, and consumers can turn waste into revenues and profits: not only through reusing, remanufacturing, and recycling but also through achieving a competitive edge.

4.3　　How Different Stakeholders Respond

We are all part of the challenge and the solution. We make and consume goods, but we also determine where they go at the end of the use cycle. But it is not only the product but also the undesired effects during making and distributing, such as GHG emissions, which need to be part of circular thinking.

In 2016, the *New York Times* wrote that, for decades, global warming created by human emissions caused ice to melt and ocean water to expand. "Now, those warnings are no longer theoretical: the inundation of the coast has begun. The sea has crept up to the point that a high tide and a brisk wind are all it takes to send water pouring into streets and homes" (Gillis 2016).

Transport represents about a quarter of GHG emissions and is the main cause of urban air pollution (European Commission, Climate Action 2016a, b). Mobility drives climate change and is hence responsible for the rise of the oceans, plus droughts and natural disasters. Plastics, in the form of packaging or as parts in product, pollute the oceans. Micro-plastics – particles of <5 mm in size – added to various consumer products are not filtered out by today's water treatment plants. Consumers also keep valuable resources in drawers and cellars; for example, valuable metals are stored in out-of-use smartphones. All over the world, governments are motivated to invest in environmental measures. Shortage or an inadequate supply of necessities, such as clean water, air, and energy, could lead to societal upheavals, even wars.

Economic wealth is also threatened by environmental pollution and climate change. According to the report "The Cost of Inaction" published by *The Economist*, the value at risk to manageable assets from climate change is \$4.2 trillion. A warming of 6 °C (approx. 40 °F) equals value losses worth \$43 trillion, that's 30% of the entire stock of the world's manageable assets (The Economist 2015). This does not take into consideration any diminished quality of life and attractiveness of sites and countries. The latter will play a major part in the ability to draw talents and investments. Not only current assets but also future potential is at stake.

Companies need to raise awareness among their employees regarding recycling of products in the office. Each year, over 375 million empty ink and toner cartridges end up in landfills or in incinerators. That's approximately 70% of all ink cartridges and 50% of all toner cartridges, though up to 97% of the materials that make up a printer cartridge can be recycled or reused. In detail, 100,000 cartridges contain 9599 kg of aluminum and 40 t of plastic (A greener refill 2016).

Subsidizing the collection of cartridges might help, since turning to an eco-friendlier product or process is often a matter of price.

All of us have to apply circular thinking to all our decisions. We need to review the content and production of all products and analyze all production processes. Policymakers need to ask themselves, with every decision they make, how it effects the environment, resources, and society. The public and private sectors have to collaborate to understand the consequences of their doing. Academia has to support the effort with research projects. Consumers have to be aware that they actively influence the supply chain with every purchase decision.

However, we need to be mindful that the awareness and urgency felt by governments, businesses, and consumers vary between developed, emerging, and developing countries. Where strong growth is still needed, sustainability might not yet be on the agenda. Global companies and international organizations are therefore needed to lead by example and educate governments, businesses, consumers, and citizens globally. This way, developing nations may be able to leapfrog straight to more sustainable systems.

The CE relies on businesses and policymakers to act. Business needs to support the design and production of eco-friendly products and manufacturing. Governments need to promote renewable energy and gradually ban pollutants and certain plastic-based products, such as plastic bags. Research has to be focused on ways to reduce/eliminate waste in all forms and reintegrate out-of-use products and other materials in subsequent supply and value chains.

Although the big changes require decision-makers in government and business to make bold moves, we can contribute our share. The challenge is enormous and time is running out. All of us are needed to contribute to the solutions in addition to electing and supporting responsible leaders.

4.3.1   Companies

We need businesses which apply responsible practices, responsible businesses which are not only guided by customer expectations and consumer needs and wants, but which also comply with regulations and lead by example. We need private sector leaders who drive sustainable strategies, innovation, products, initiatives, programs, and practices across their organization, the industry, and economy. There are many examples that illustrate ways of combining economic success with responsible ecosystem leadership.

Brooks, a manufacturer of running shoes and apparel, is among the companies that think and act in a circular way already. Brooks was able to reduce the amount of materials and intermediate stages of the production process and, hence, the consumption of energy. Brooks also created a biodegradable midsole material (named BioMoGo). By using BioMoGo in its shoes, Brooks will be able to save nearly 30 million pounds of landfill waste over a 20-year period. They also state that their shoe is made from 75% recycled materials (Brooks Blog 2015). Scrap rubber is shredded and reused. That makes Brooks a pioneer in implementing a circular supply and value chain.

O'Neill took up the battle against plastic scrap in the oceans. In April 2016, the developer and manufacturer of surf clothing announced its Blue Collection, a range of clothing made from recycled beach plastic (Chapman 2016).

Waste reduction is on *Unilever*'s strategic agenda. The company is a member of the CE100 program established by the Ellen MacArthur Foundation (Ellen MacArthur Foundation 2016). By applying new strategies to its production cycle, Unilever reduced a total of 85% waste per ton of production since 2008. In addition, the company is committed to reducing packaging waste. According to Unilever, they have cut their waste footprint by 12% per consumer use in 2014 versus their 2010 baseline (Unilever 2016). Also, lighter packaging was designed – making shipping less costly and reducing energy and emissions in transport. Unilever drives the reuse of materials, either on site or through trading into other industrial supply chains; organics are composted. According to GreenBiz, reducing waste in Unilever factories has saved the company $226 million (Marks & Spencer 2016). In regard to sustainable sourcing from farms and forests, Unilever aims to receive 100% of their agricultural raw materials sustainably by 2020. Unilever wants to reduce deforestation too. The company announced that over 90% of their globally traded palm oil is now covered by "no deforestation" pledges (Unilever 2016).

With similar goals in a different industry, *Rolls-Royce* is aiming to send zero waste to landfills by 2020. At this date, the British manufacturer claims 11% progress. Driven by the "Revert" program, manufacturing sites around the world are requested to recover waste metals, so that these can be melted and turned into new aerospace alloys. In the future, the manufacturer aims to integrate more business partners and customers into the program (Rolls Royce 2016).

In the UK, *Marks & Spencer* wants to be a pioneer in sustainable production and consumption. According to the company's estimates, the Marks & Spencer customer base generates over 400,000 t of materials with just over half likely to be reused or recycled. The company committed to send zero waste to landfill from its own operations and construction activities in the UK and the Republic of Ireland. In 2012, the management announced to have achieved this goal and has asked their suppliers to follow their ambition and join the program (Marks & Spencer 2016).

Also in the UK, *Banton Frameworks* offers customized eyewear according to material preferences, job, and even skin tone. Customers can choose from seven frame designs, five materials, three color options, six temple color finishes, and two leather earpiece colors. Though the lenses are not reusable, the frames are. This allows the company to replace only the lenses if they need to be adapted to the customer's eyesight, or they can be refurbished for subsequent use.

A company that delivers both customization and eco-friendliness is *Eredetti Sunglasses* in Rexburg, Idaho. Their wood frame eyewear is made with wood or bamboo. By its own account, the company strives to seek out eco-friendly sources and products at all levels, from the microfiber bags, which are made from recycled felt, to the recycled cardboard boxes used for shipping. Customers can choose from five models and various finishes.

Patagonia, a well-known outdoor clothing company, launched the "Common Threads Initiative" and started focusing on repairing, reusing, and recycling used gear. Today, the company employs 45 full-time repair technicians at their service center in Reno, Nevada. It's the largest repair facility in North America – completing about 40,000 repairs per year.

Polli Boat is the world's first trimaran inspired by the 3R practices: Reduce, Reuse, and Recycle. The boat is designed and engineered by *Miniwiz*, a pioneer in the circular space, and made of recycled materials and a newly developed green material called wood-plastic composite (WPC). The boat uses waste canvas as decking, and its renewable energy systems, powered by both solar and wind, are interchangeable depending on weather conditions.

Dell collects damaged and used products to reprocess plastics to be used in the manufacture of new products. *Caterpillar* brings hundreds of refurbished parts back into the marketplace every year, while *Rolls-Royce* is remanufacturing marine engines. *Renault* takes the complete life cycle of electric vehicle batteries, including recycling, into account[3]; the goal is to recycle all the elements contained inside the battery.

Waste can be used to generate fuel: unsuitable meat products, scraps of plants, and industrial emissions are being used. In New Mexico, car manufacturers *Audi* and *Joule Unlimited*, a Massachusetts-based pioneer of CO_2-to-fuel production platforms, are looking for approaches of directly making ethanol from sunlight and CO_2. With the aid of sunlight, CO_2, and water, microorganisms produce fuel molecules in their cells, which form the basis for Audi's synthetic fuels. The gas is taken from industrial plants. Instead of fresh water, the partners use wastewater. Additionally, the synthetic fuel generates less pollutants during combustion than

---

[3]Watch Renault's video on YouTube: https://www.youtube.com/watch?v=4zSWTT0qCzQ

fossil-based fuel. It contains neither olefins nor aromatic hydrocarbons (Audi Encounter 2014).

Neste Oil has pursued a different biodiesel method. In 2011, the group started to operate biodiesel plants at the port of Rotterdam, Netherlands, and at Tuas industrial zone, Singapore. The tanks at the oil refineries are filled with unsuitable animal fat, remnants produced by the fish industry, and used frying oil from canteen kitchens. Henceforth, it is producing 800,000 million to 1 billion tons of biodiesel annually (Reuter et al. 2013).

Lexmark is a provider of toner cartridges. The company's customers in the USA return between 45 and 50% of the toner cartridges. In 2014, Lexmark recycled or reused more than 10,000 metric tons of plastic, metals, and packaging and was able to materially recycle or reuse 100% of the reclaimed plastic (Lexmark 2016). Most major companies like *Canon*, *HP*, *Brother*, and *Xerox* along with office suppliers *Staples* and *Office Depot* have implemented return programs; most of them even pay for the shipping. In Australia, leftover toner which accounts for on average 13% per cartridge was given a new purpose as asphalt. The eco-friendly mix contains recycled printer toner and was first used in Melbourne in 2013. According to *Close the Loop*, a local cartridge recycling company, the toner is blended with recycled oil, hence 40% more energy efficient than the manufacture of standard bitumen, with a relative saving of 270 kg of CO_2 emissions per ton (Tan 2015).

Car manufacturer *Volkswagen* (VW) has established a process that allows them to recycle 95% of a vehicle's components (by weight). In the Golf, for example, secondary raw materials make up 40% of each vehicle by weight; 95% of those materials are metals, the company states. The company has also codeveloped the so-called VW SiCon process. Thanks to this method, shredder residues can be turned into new raw materials instead of being sent to landfills. In 2006, this process was awarded the European Business Award for the Environment and the Environmental Award of the Federation of German Industry (BDI). This is how VW describes the process: "The end-of-life vehicle is first drained of fluids, then a range of components which are capable of being recycled and used in new products are removed. Any parts not subject to statutory end of life dismantling requirements, or which cannot be cost effectively remanufactured into replacement parts, are put through the shredder. Unlike conventional recycling systems, the VW SiCon recycling process is also capable of obtaining secondary raw materials from nonmetallic shredder residues" (Volkswagen AG 2016). That way, the car turns into a source of raw materials.

As noted previously, the Sustainable Packaging Consortium aims to reduce weight of packaging in order to reduce shipping costs and to foster sustainability. When it comes to the actual material, companies are also offering packaging matter that contains recycled contents only. One of them is *Salazar Packaging* in Illinois. Their shipping boxes, die cut mailers, paperboard cartons, rigid wall retail boxes, partitions, dividers, and inserts are all made from recycled and recyclable content materials.

Another industrial leader in regard to sustainability is the Canadian company *Strathcona Paper*. They designed a paperboard that is EcoLogo, FSC, and

Rainforest Alliance certified at the same time. Their food packaging has passed all Health Canada, Canadian Food Inspection Agency, and FDA requirements for direct food contact. Strathcona Paper claims to be the only recycled boxboard manufacturer to have received direct food contact accreditation by a Canadian authority.

Global furniture retailer *Ikea* is also planning to become greener in regards to its packaging. The company is planning to use packaging made with mushrooms as an eco-friendly replacement for polystyrene. Using the biodegradable mycelium, "fungi packaging" is one part of its effort to reduce waste and increase recycling. The fungus grows in a mass of branched fibers, making it ideally suited to being made into packaging material. The American company *Ecovative* developed this product, by letting the mycelium grow around clean agricultural waste, such as corn stalks or husks (Gosden 2016). *Apple* is going in a similar direction. In April 2015, it purchased more than 36,000 acres of forest in Maine and North Carolina, with the goal to sustainably harvest and use the material to produce packaging (Lyons Hardcastle 2015).

The World Economic Forum report, "Beyond Supply Chains: Empowering Responsible Value Chains," has identified a set of 31 proven responsible supply chain practices. Analysis of these practices has shown that companies applying responsible practices can "increase revenue by up to 20% for responsible products, reduce supply chain costs from 9–16% and increase brand value by 15–30%" (World Economic Forum 2015).

4.3.2 Governments

The action of the private sector needs to be flanked by policies and regulation. Policymakers must discuss, prepare, and launch planet-friendly regulations, initiatives, and practices on country and city levels. Sweden, for example, announced in 2014 that more than 99% of all household waste was recycled (Sweden.se 2016). In fact, Sweden has become so good at recycling waste that the country needs to import trash from other European countries to keep the waste-to-energy (WTE) plants running.

France is trying to take the lead in reducing GHG emissions in the transport sector. By launching Decree No. 2011–1336 in October 2011, the French government required transport companies to transmit information on the quantity of carbon dioxide emitted during transport. The provision applies to several transport modes (rail, road, air, maritime, and waterways) and covers CO_2 emissions from source to wheel. The decree entered into force on October 1, 2013 (Ministère de l'Environnement, de l'Energie et de la Mer 2016). However, since noncompliance with the decree will not be sanctioned, the provision is not taken seriously by most transport companies. Most companies don't even know about this rule (Beck 2013).

The EU issued regulations for a carbon emissions trading system (EU ETS) in 2005. Up to now, the system has covered only power stations, industrial plants, and airlines. The EU ETS works on the "cap and trade" principle. A limit (cap) is set on the total amount of certain GHG that can be emitted. The cap is reduced over time. By 2020, the EU hopes to have reduced emissions compared with 2005 from sectors covered by the EU ETS by 21%. By 2030, the Commission proposes, GHG would be 43% lower (European Commission 2016).

The Dodd-Frank Act (DFA) is an example of targeted measurements. The full name of the U.S bill is the Dodd-Frank Wall Street Reform and Consumer Protection Act. The bill contains several specialized disclosure provisions including Section 1502, which requires listed companies to disclose annually whether "any conflict minerals that are necessary to the functionality or production of a product of the person, as defined in the provision, originated in the Democratic Republic of the Congo or an adjoining country" (U.S. Securities and Exchange Commission 2016). The EU is working on passing a similar legislature. In contrast to the U.S bill, the EU suggests establishing a responsible importer certificate that aligns to OECD guidelines regarding due diligence in the supply chain. This voluntary disclosure has been criticized, along with the intention to apply this legislature to trade and quarrying of raw materials. If and when this bill might be passed remains to be seen.

Spotlight: Conflict Minerals

Conflict minerals is the term for tin, tungsten, tantalum, and gold that have been mined under dubious circumstances, either by illegal mining or under dangerous working conditions. They are used in parts of electronic equipment as well as in-seat cushions, batteries, and much more. In today's increasingly complex supply chains, it is sometimes difficult to determine exactly who the seller is. Insight is needed to avoid fines and damage to one's reputation.

Even though certificates and legal regulations are in place, a company should investigate the source of all materials thoroughly for their own protection. A report from Tulane University and the consulting firm Assent Compliance found that, despite spending $709 million on compliance efforts in 2014, 90% of the 1262 companies that filed conflict minerals reports with securities regulators said they were unable to determine whether their products were conflict-free. This is largely due to a federal appeals court ruling contradicting the Dodd-Frank Act (see Sect. 4.2.3), which states that companies could not be forced to hurt their own products by reporting that they were unable to determine if they were conflict-free.

In 2014, power plants accounted for nearly 40% of U.S CO_2 emissions. More than the emissions of every car, truck, and plane in the USA combined. Therefore, the U.S Environmental Protection Agency (EPA) is developing standards for GHG emissions from mobile and stationary sources under the Clean Air Act. It includes the Clean Power Plan, which aims to cut harmful pollution from the power sector by 32% below 2005 levels, plus cut smog- and soot-forming emissions that threaten public health by 20% by way of establishing guidelines, state-based programs for existing sources, and federal programs that establish standards of performance of new, modified, and reconstructed power plants. In August 2015, the EPA issued two proposals to further reduce emissions of methane-rich gas from municipal solid waste (MSW) landfills. The proposals require new, modified, and existing landfills to begin capturing and controlling landfill gas at emission levels nearly a third lower than current required levels.

The EPA and the NHTSA are taking steps to reduce GHG and improve fuel efficiency. Their goal is to lower fuel consumption and CO_2 emissions by 24% for commercial tractor-trailers, 16% for heavy-duty vehicles, and 16% for commercial pickup trucks (ETA 2016). Other legislations aim to facilitate the broader use of less harmful energies as well. The Energy Policy Act of 2005, for example, establishes regulations regarding loan guarantees for entities that develop or use innovative technologies that avoid the production of GHG. Another provision of that act increases the amount of biofuel that must be mixed with gasoline sold in the USA. In addition to that plan to push renewable energy, the U.S Department of Energy launched the SunShot Initiative in 2011, with the goal of cutting the total cost of photovoltaic solar energy systems 75% by 2020.

Consumer electronics products, in particular, are outdated in no time. E-waste, therefore, is becoming both a major problem and a major industry. In the USA, citizens scrap about 400 million units per year of consumer electronics. Only about 40% of them are recycled. The EPA has found that about 70% of the toxic metals, such as lead, in municipal solid waste landfills came from discarded electronic items. The agency states that recycling one million laptops saves the energy equivalent of the electricity used by more than 3500 US homes in a year.

The government addressed e-waste with their National Strategy for Electronics Stewardship (NSES), which was initiated in 2011. It aims to lay the groundwork for improving the design of electronic products, enhance management of used or discarded electronics, and reduce harm from U.S exports of e-waste. It also promotes donation of used electronic and lists state-owned drop-off sites (Environmental Protection Agency 2016a, b). Websites like *earth991.com*, *call2recycle.org* (for cell phones and batteries), and *ecyclingcentral.com* provide similar information. *GreenerGadgets.org* lists nationwide manufacturer, retailer, and certified eCycling locations, in addition to providing advice for purchasing sustainable products. Most of the major retailers, including *Staples* and *Best Buy*, offer drop-off sites at their branches; manufacturers, including *Dell*, *Samsung,* and *LG*, offer mail-in recycling options; *Sony* will even come to pick up consumer TVs.

The federal government operates prison-based recycling plants. *UNICOR*, a company that is wholly owned by the federal Department of Justice, operates electronics recycling in seven federal prisons. As of 2010, UNICOR operated 103 factories in 73 federal prisons, employing about 17,000 people. However, reviews found that neither health nor safety issues were sufficiently addressed (U.S. Department of Justice 2010). In 2015, the EPA released figures that indicated success in the recycling of e-waste: consumer electronics recycling went up from 30.6% in 2012 to 40.4% in 2013.

In the EU, policymakers stress the importance of less waste and more circularity. The European Commission recently presented figures and initiated a new agenda:

the Circular Economy Package includes a recycling ratio of 65% of municipal waste and of 75% of packaging waste by 2030. The legislature wants to stimulate Europe's transition towards a circular economy by "closing the loop of product life cycles through greater recycling and re-use, and bring benefits for both the environment and the economy" (European Commission 2016).

Cities are taking action as well. San Francisco banned the sale of plastic water bottles in December 2015. Until 2020, the city will phase out the sales of plastic water bottles that hold 21 ounces or less in public spaces. It is believed that plastic pollution is one of the greatest burdens to the environment. Experts estimate that 50% of the plastic on this planet is used only once before being thrown away. In July 2015, Hawaii launched a similar ban and has stopped grocery stores from distributing plastic bags.

Micro-plastics, in particular, could prove to be harmful not only for the environment but also for us. The tiny plastic beads and the chemical additives they contain will stay on the agenda of political leaders for some time. International organizations and national governments have started to work on the threat. Various US states have addressed the topic the NY State Assembly passed the Microbead-Free Waters Act in early 2015, which will prohibit the distribution and sale of cosmetic products containing plastic microbeads <5 mm in size in general. California adopted a legislation banning tiny plastic beads in personal care products in 2015, allowing a 5-year phasing-out period. The Canadian government has announced a general ban on micro-plastics as well. Micro-plastics will be added to the list of toxic substances. The European Environment Council banned rinse-off cosmetics containing microbeads from carrying the EU Ecolabel.

In 2011, China consumed more raw materials than the 34 countries of the Organization for Economic Co-operation and Development (OECD) combined: 25.2 billion tons. In 2014, China generated 3.2 billion tons of industrial solid waste, 2 billion tons of which was recovered by recycling, composting, incinerating, or reusing. By 2025, China is expected to produce almost one-quarter of the world's municipal solid waste (Mathews and Tan 2016).

Chinese policymakers have become more conscious of environmental issues and the harmful effects to life. In early 2016, China announced the closure of 4300 small and inefficient coal mines over the next 3 years. Chinese public leadership aims to reduce the share of coal in the overall energy mix as part of efforts to cut smog and GHG emissions. Up to now, coal consumption has been estimated to amount to up to 64.4% of China's total energy mix (Reuters 2016).

The Chinese government has taken concrete steps towards sustainability. The China Circular Economy Promotion Law came into force in January 2009. Its purpose was to facilitate the circular economy, raise the resource utilization rate, protect and improve environment, and foster sustained development (Invest in China 2016). As a result, eco-cities and industrial clusters came into existence. The Suzhou New District is one example. The 52 km² region for technological and industrial development near Shanghai is home to around 4000 manufacturing firms. There, for example, manufacturers of printed circuit boards use copper that is recovered from waste from elsewhere in the park, rather than using virgin copper produced by mining firms (Mathews and Tan 2016).

In 2015, the G7 committed to ambitious action to advance the efficient use of natural resources throughout their life cycle.[4] "For the first time, the leaders of the G7 have officially recognized the importance of the link between materials recovery and the global economy, and established the G7 Alliance on Resource Efficiency," said Mathy Stanislaus, assistant administrator for EPA's Office of Solid Waste and Emergency Response (Environmental Protection Agency 2015). The G7 Alliance on Resource Efficiency aims to provide an exchange forum for G7 members on how to foster and communicate the circular economy.

4.3.3   Consumers and Citizens

Reutilization of products beyond their initial life cycle requires the support of consumers. Consumers decide whether or not to stockpile appliances at home instead of returning and reinjecting out-of-use products into the closed-loop system. The consumer has to be willing to recycle plastics, paper, cans, and household waste.

The Stanford University's "Alumni" states that between 60 and 80% of products labeled as recyclables are actually recycled, depending on the bin and on the city's recycling system (Cima 2016). According to the EPA, only 70.6% of steel cans sold and 67% of newspapers and mechanical papers have actually been recycled in 2013. Almost half of soda, beer, and aluminum cans (55.1%) were taken to recycling facilities. Other products achieved lower recycling ratios: consumer electronics, 40.4%; PET bottles and jars, 31.3%; and white plastic bottles (HDPE), 28.2% (Environmental Protection Agency 2016a, b).

Often, consumers don't seem to be aware of their degree of influence. Consumers buy clothes that are cheaply made and which they have to throw away after a few washes. It would be better to turn instead to more expensive fabrics with a higher quality, ensuring a longer use cycle. Consumers who purchase products produced in an eco-friendly way and in accordance with work regulations often question whether or not they really influence the supply chain in a positive way. The doubt is rooted in the lack of transparency. Does the worker in a factory far away actually receive higher pay, or does it go to the manufacturer instead? If the cotton used to make a shirt is organically sourced, what about the dye? This lack of transparency leaves consumers at a loss. Nevertheless, consumers are responsible for their decisions. In our increasingly digitized world, consumers should be more comfortable that full transparency will come and responsible behavior and choices influence and pay in the end.

---

[4]Group of seven major advanced economies includes the USA, Canada, France, Germany, Italy, the UK, and Japan.

Consumers are becoming increasingly open to paying for better and more sustainable products. According to research from Cone Communications conducted in 2013, 71% of respondents consider the environment when they shop, which is up 5% from 2008. However, the majority of respondents (85%) said they do not feel sufficiently educated concerning the use and disposal of green products. The survey also indicated that consumers value honesty. Seven in ten respondents (69%) said that they understand that no company is environmentally perfect and are fine as long as this deficit is openly communicated (Shearman 2013).

Through digitalization, customers and other stakeholders have a different standing and a broader range of possibilities than a few decades ago. Through the Internet, they receive and can acquire much more information. They are having a say towards certain actions, and they can impact trends or decisions by stating their opinion online. Thus, all of us can influence the supply chain to a certain degree. Customers and consumers trigger supply chains with their orders and determine which products are made.

Consumption patterns play a major part in establishing and asserting guidelines. It provides policymakers with information on how people deal with products or residual material. Today, product portfolios are frequently changed. Manufacturers feel they have to respond quickly to changing customer requests. Consumers have to be aware too that demand often determines production conditions, the way products are shipped or delivered. For instance, certain price levels are connected to slim margins for retailer as well as low profits for manufacturers and low wages for workers. Most of the time, the price determines the way of manufacturing and, along with these, working and living conditions. These sometimes have a disastrous impact on the health and well-being of all participants along the production process. Buying responsible products diminishes negative impacts. For example, when buying gold or products that contain gold, such as cell phones, consumers should be mindful that extracting gold causes an environmental burden: for one ounce of gold, up to 91 t of rock needs to be moved (Ro 2013).

In some parts of the world, mercury is still used to separate the valuable metal from ore, often in unsafe and environmentally damaging ways. The United Nations Industrial Development Organization estimates that 1000 t of mercury is released into the air, soil, and water each year by artisanal and small-scale miners (ASM) (International Council on Mining & Metals et al. 2009). In most developed countries, cyanide, in the form of a very dilute sodium cyanide solution, is used to dissolve and separate gold from ore. Water and a cyanide solution is applied to crushed rocks, which is then ground into a mud-like substance. In settling tanks, the more solid pulp settles on the ground, and the water drains to another area. Because cyanide is toxic in large doses, it is strictly regulated in most jurisdictions worldwide to protect people, animals, and the aquatic environment. After severe spills, the mining companies also created the International Cyanide Management Code, which includes minimizing the amount of cyanide used, creating measures to protect surface and groundwater, and a design and operation of systems to reduce cyanide levels in effluent.

Consumers also have a choice. *Fairmined* is a label that certifies gold from responsible artisanal and small-scale mining organizations. The Alliance for Responsible Mining (ARM) certifies mines based on Fairmined standards. They include fair wages for the miner, safe and reduced handling of chemicals or chemical-free extraction, protection of water supplies, no child labor, and no link to any conflict situation (Fairminded 2016).

Endomines, a Finnish mining and exploration company, has taken responsibility and eco-friendliness even further: in one of its mines in Pampalo, Finland, the company does not extract gold from other substances by leaching with sodium cyanide, but gold concentrates are refined by smelting (Endomines 2016).

Recycling poses a much safer and more environmentally friendly way of obtaining gold. But the amount of recycled gold in the market fluctuates with the market price for gold and the global economic situation. In 2009, with high gold prices and a stagnant global economy, recycled gold reached an all-time high of 42% of the total gold supply. 2014 saw the lowest level of recycling since 2007 with 26% of total supply. According to the Boston Consulting Group, the recycled-gold supply will increase over time: over the long run by 4% annually (The Boston Consulting Group 2015).

The consumer has various ways of participating in the circular supply chain and economy. Books, clothes, and used consumer electronic devices can be donated, lent, or sold. Recycling yards, Internet-based services, and retailers stand ready to take recyclables and sometimes even pay for shipping. Of course, this has an economic reason: recycling pays.

Consumers should be aware and pay attention to the working conditions. Human rights advocates note that some companies in emerging countries maintain slave-like labor forces. According to the 2014 Walk Free Foundation index, almost 36 million people live in "modern-day slavery" conditions, due to human trafficking, forced labor (including of children), and hazardous work environments (Walk Free Foundation 2014).

Certifications inform about the origin of certain foods or fibers. The National Organic Program (NOP) certifies organic growers and handlers who prove to comply with organic regulations. The Global Organic Textile Standard (GOTS) is viewed as an alternative certification for processed textiles. GOTS is a voluntary international standard for the processing of textiles containing organic fiber, which addresses the entire post-harvest processing, including spinning, knitting, weaving, dyeing, and manufacturing, of apparel and textile products made with organic fiber. Candidates have to prove that their product contains at least 70% organically sourced fibers. They are not allowed to use certain toxic or environmentally harmful dyes or chemicals and need to adhere to water treatment standards. In addition, basic social standards have to be kept, including no child labor, living wages, hygienic working conditions, as well as freedom of assembly and association.

Coming back to the supply chain, consumers can put pressure on companies to act more sustainably; to adhere to standards of living wages, hygiene, and safety; and to use organic or fairly traded products by educating themselves about products and purchase accordingly. They must look for certain labels. Consumers can educate themselves by different means. *The Ethical Consumer*, a magazine and website, provides guidelines for purchases. The EPA has put up a website to inform consumers (US Environmental Protection Agency 2016). Many manufacturers provide the possibility to scan barcodes and QR codes on the products with smartphone apps to obtain information.

4.3.4   International Stakeholders

Sustainable and responsible actions are a powerful way to differentiate oneself and stand out, as a country, industry, organization, or individual. Commitments and communications to stakeholders need to be credible. Credibility can be achieved by clearly defined visions, actions, and programs transparently followed through and demonstrated by public key performance indicators (KPI) or scorecards. Audits conducted by impartial third parties add to the credibility.

Working closely with organizations, such as the Ellen MacArthur Foundation, International Transport Forum at the OECD, World Bank, World Business Council for Sustainable Development (WBCSD), UN Global Compact, World Economic Forum, and others, helps to define the most responsible way forward and to find like-minded allies and supporters. Numerous organizations have been established to support and enable ethical conduct. Some are listed and described in the following.

The *ILO* was founded in 1919 after World War I, "To pursue a vision based on the premise that universal, lasting peace can be established only if it is based on social justice" (ILO 2016). In 1946, the ILO became the first specialized agency of the United Nations. The organization focuses on promoting rights at work, encouraging decent employment opportunities, and enhancing social protection. One goal is to eliminate child labor.

ETI is short for Ethical Trading Initiative. This alliance of companies, trade unions, and nongovernmental organizations (NGO) promotes workers' rights including trying to stop exploitation and discrimination and securing free, safe, hygienic, and equal working conditions. The organization also fights child labor. The ETI also promotes the payment of living wages and reasonable working hours. Members committing to the code aim at a more just and respectful world, while enhancing brand integrity and establishing a more resilient supply chain (Ethical Trading Initiative 2016).

The *Fairtrade Foundation* was established in the UK in 1992. Fairtrade, the trade in products from certified sustainably sourced producers, grew increasingly popular in the following years. In 2013, global retail sales reached 4.4 billion pound sterling (approx. $6.2 billion) (Smithers 2014). In 1998, the Institute for Agricultural Trade Policy (IATP) founded *Fair Trade USA*. The nonprofit organization shares the same goals as the British Foundation: to ensure that farmers work in safe conditions, to receive a harvest price that enables them to make a living, and to protect the environment. In 2011, Fair Trade USA parted from the Fairtrade Foundation, of which it was a member. Paul Rice, President and CEO of Fair

Trade USA, said that he wanted "to significantly increase the effectiveness and reach of the Fair Trade model. (Thus) Fair Trade USA is embarking on a new vision, Fair Trade for All, aimed at doubling the impact of Fair Trade by 2015 by innovating the model, strengthening farming communities and igniting consumer involvement" (Rice 2012). According to Fair Trade USA, U.S retail sales of Fair Trade products reached $1.5 billion in 2011. The organization currently partners with over 1000 companies and brands in over 80 countries. Certified commodities include coffee, tea, cocoa, sugar, spices, honey, produce, grains, wine and spirits, flowers, apparel, home goods, and body care (Fair Trade ISA 2016).

Schemes and standards are an important tool to drive and measure actions. Organizations which wish to demonstrate their eco-performance can leverage ISO 14001 and the Eco-Management and Audit Scheme (EMAS) of the EU. The ISO 14001 by the International Standardization Organization (ISO) requires businesses to establish and document environmental management systems. They also have to set and follow through specific environmental objectives.

Once ISO 14001 certified, organizations located in the EU can apply for another voluntary code of management: EMS allows all types of organizations to improve their environmental performance. The scheme aims at recognizing and rewarding those organizations that go beyond minimum legal compliance and continuously improve their eco-performance. The scheme requires companies to prove that they have established an effective environmental management system (EMS), which ensures that an environmental policy is in place and objectives, targets and programs are set to continuously improve environmental performance. In the USA, similar voluntary codes alongside the ISO norm do exist but are not used widely. The EPA developed an environmental management system (EMS) that is a set of processes and practices that aims to support the organization in reducing its environmental impacts and increase its operating efficiency.

4.4 The Paradigm Shift in Supply Chain Management

Since the supply chain has to prove its effectiveness in securing economic growth, wealth, and sustainability, the pressure on the industry, and CSCOs, is high. The supply chain has become both the backbone of global production networks and a primary provider of solutions for supply and disposal flows in urban and rural areas, nationally and globally. In the interest of people and the planet, the supply chain industry should help to establish the CE&S, with the circular supply chain (CSC) at its core.

The CSC aims to replace the linear value chain, which is neither regenerative nor as value adding as the closed-loop model. The linear economy and society do not carry the total cost of their actions. The cost to mitigate, repair, or deal with the environmental and other damages caused by linear behavior is often born by governments and not by the economy which has caused the damage.

However, not only businesses need to rethink and act. Flows along the supply chain can only help to create circularity provided all stakeholders support the effort and are working together. Establishing a circular supply chain in a degenerative system will not work. We are collectively responsible for the cleanliness of the air, rivers, lakes, ocean and the ground, and the future availability of resources. We decide how we treat the planet, people, and future generations.

In one of my previous books, I introduced the logistics paradigm: a framework to establish successful logistics businesses. Reviewing this paradigm through the circular supply chain lens unveils the need to drive the thinking further. Each dimension of the model and the core qualities for successful logistics management need to be enhanced to establish a solid and sustainable circular supply chain paradigm.

Core qualities of the logistics paradigm are process mastery and relations management. The framework recommends defining vision, strategies, concepts, and structures by falling back on these two qualities. The recommendations derived from the core qualities are expressed in regard to the following four dimensions: value chain thinking, life cycle view, customer focus, and societal contribution (Lehmacher 2013).

Logistics service providers, in particular, have to practice *value chain thinking* – that is, understanding and designing their own value chains, in addition to identifying and cocreating those of current and future partners and customers. Logisticians have to think in value chains. Hence, they need to develop knowledge on the characteristics and structures of certain flows of goods. Knowledge about specific vertical supply chains, such as electronics, food, or chemicals, is the basis on which to design and establish top-class value and supply chains. With knowledge and understanding in tow, logisticians are able to create efficient and durable end-to-end models which create value for suppliers, customers, and themselves.

A closer inspection of the *life cycle view* uncovers the need for reuse of out-of-use products, parts, and materials. For logistics companies, this is a field of new business opportunities. Entering the field of repurposing requires specific knowledge in material science and the production of the respective products. Building up this knowledge requires collaboration between the respective manufacturer and supply chain service provider in terms of access to product information. Product trainings are one form of knowledge transfer. The continuous involvement of the logistics service provider from the product design and new product introduction (NPI) process through manufacturing to distribution, after-sales, and repurposing is the ideal form of life cycle collaboration.

The third dimension, *customer focus*, takes the strategy design from the customer angle. The goal is to optimize customers' perceptions by monitoring and measuring the quality of experience and perception of all participating parties, including customers' customers. Martin Andersen, managing director at *Amazon Germany*, has explained to *VerkehrsRundschau*, a magazine for the transport sector, their customer-centric approach. In order to optimize customer service and customer experience in parallel, Amazon is working its way backwards. Officers take on the role of customer and report back regarding at which stages of the process they experienced potential for improvement. For example, if they felt a need for faster delivery, Amazon would check their logistics processes in order to shorten flow processes (VerkehrsRundschau 2014).

The fourth dimension goes beyond customer focus to *societal contribution*. The question here is how logistics service providers can contribute to an effective supply chain ecosystem, and moreover, do direct and indirect stakeholders, from government to media and civil society, consider this contribution to be valuable? Logistics and supply chain service providers can position themselves as pioneers at the center of sustainable behavior and serve as advisors to peers and other industries.

By analyzing these four dimensions, companies in the logistics industry and their corporate customers and other stakeholders will gain insights on inherent opportunities and threats to the supply chain ecosystem. However, in light of the CSC, our thinking and acting have to move beyond the boundaries of this paradigm. While the logistics paradigm focuses on long-term economic success of a business for the benefit of society, the CSC aims to be more supportive and beneficial to society by being instrumental to the transition from the linear to the circular model.

Change in mindset, new knowledge, and comprehensive and effective programs and measures are all necessary in order to live up to this ambition. To illustrate the additional mile, the logistics paradigm needs to go to be extended in all four dimensions: value chain thinking needs to become partner integration; life cycle view transforms into cradle-to-cradle approach; customer collaboration takes the place of customer focus; and societal contribution expands into global stewardship.

The core qualities to create the new world of supply chains are knowledge management and cocreation capacity. Creating the circular economy and society requires cocreation across all stakeholder groups, based on new knowledge which responds positively to the increasing challenges of today's world. In our modern, global, hyperconnected, and interdependent world, stakeholders have yet to understand the challenges in full scope and to create concrete solutions for realizing the CE (Fig. 4.1).

The extension of the logistics paradigm does not make its current dimensions obsolete. However, new approaches and abilities have become an important additional step to the original framework. Extending the business-driven paradigm results in the circular supply chain paradigm, outlining the set of directions pointing towards the CSC and the CE&S. Moving beyond the logistics paradigm includes several important steps towards creating the CSC.

Modern technology allows seamless *partner integration*. This integration increases the visibility along the supply chain. Visibility is needed to manage the supply chain responsibly and to track products throughout the entire use cycle and beyond. Integration also allows companies to efficiently run partner programs to recover materials and out-of-use products for further utilization.

Life cycle view must transform into a *cradle-to-cradle approach*. Once materials, parts, and out-of-use products are recovered, they need to be repurposed. This can only happen if the design considered a subsequent use cycle. The philosophy and approach support proper design to ensure the endless utilization of

Transitioning to the Regenerative World

Moving from a more value-driven logistics thinking and framework to enabling and shaping a sustainable supply chain ecosystem

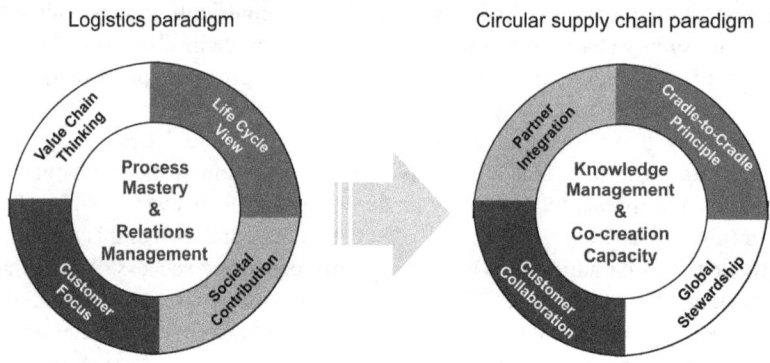

Particular importance is placed on global responsibility and the constructive interactions between stakeholders towards a zero waste and zero emission world

Fig. 4.1 Transitioning to the regenerative world

resources – from cradle to cradle. CSCOs should contribute to the health of the entire supply chain ecosystem and support the cradle-to-cradle concept.

Customers are hard to integrate. However, with new technologies come new ways of *customer collaboration*, engaging business customers and consumers. As customers and their behavior are better known, targeted collaborative efforts can be designed to help reduce energy utilization, carbon emissions, and waste. *Levi's*, for example, is asking customers to recycle their jeans to reprocess the fibers. According to Levi's, we could save 30.6 million metric tons of carbon dioxide and equivalent emissions if everyone in the USA recycled their clothing and textiles for 1 year (Triple Pundit 2015).

Finally, CE requires not only societal contribution, but *global stewardship*, meaning leaders assume an active role in protecting the planet and people. Leaders in the private and public sector first need to understand the effects of their own decisions on the environment and society in their area and beyond to be able to adjust their business practices. Once new practices show effect, leaders need to showcase responsible practices globally.

4.5 Roadmap Back to the Regenerative World

The concept of the CE&S and CSC envisions an economy with governments, organizations, and a population that does not need to fear an unhealthy, unjust, and unsafe future. Most of us will need to change behavior. Chief supply chain officers have to overcome their current manner of planning and implementing supply chains, which primarily focuses on optimizing processes and capacities to reduce costs while ensuring the expected level of performance. In the future, they have to establish customer-centric and society-driven systems, aka the circular supply and value chain, which consider not only optimal costs at a given performance level but also the potential impact on environment and society.

An understanding of interdisciplinary and global correlations is essential. Thinking far beyond company borders will be a key requirement. Processes do not stop at the end of a use cycle. Some processes are responsible, some are toxic. CSCOs, together with their colleagues, need to find ways to find subsequent use for products, parts, and materials. CSCOs need to help to reduce carbon emissions and fight child labor and corruption.

The CSC claims high standards regarding zero waste and zero emissions. Collaboration is the order of the day for companies, product designer and manager, supplier, business customer, consumers, and waste management and recycling companies. Close collaboration requires a high degree of openness, determination, and creative thinking and the capability to foster target-orientated cocreation.

CSCOs carry a higher share of responsibility. Indirectly and directly, they are responsible for the economic success of their organizations and the well-being of mankind and the planet.

The cradle-to-cradle concept seems to be the tool of choice regarding circular thinking and behavior. It imitates nature's system of closed loops – nothing goes to waste, and everything is reused. The phrase cradle to cradle was popularized by Martin Braungart, Professor at Technical University Munich. In 2002, Braungart and William McDonough published the book *Cradle to Cradle: Remaking the Way We Make Things*, which presented efficient and essentially waste-free systems. By following cradle-to-cradle principles, companies will be able to position themselves as models for farsighted, eco-friendly, resource-saving, and socially responsible activity.

It is important that business set the standards for responsible performance. The CSCO of tomorrow needs to be able to refer to responsible strategies, ethics, and values established by the organization. The intention to implement cradle to cradle and the CSC will drive research and development, the design of products, operating models, and investment plans. The roadmap towards the CSC can be derived from the answers to key questions in the four areas of the circular supply chain paradigm. Figure 4.2 depicts some potential questions in this respect.

At the core, the roadmap towards implementing circular thinking and behavior will have to include the following steps:

The Circular Supply Chain Paradigm

At the core of the framework are knowledge management and co-creation capacity

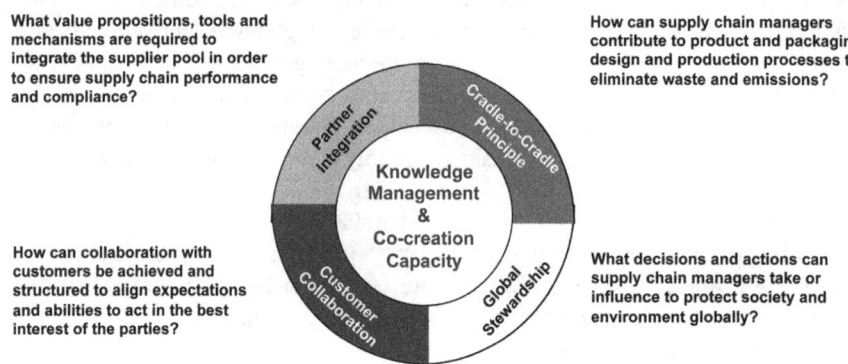

What value propositions, tools and mechanisms are required to integrate the supplier pool in order to ensure supply chain performance and compliance?

How can supply chain managers contribute to product and packaging design and production processes to eliminate waste and emissions?

How can collaboration with customers be achieved and structured to align expectations and abilities to act in the best interest of the parties?

What decisions and actions can supply chain managers take or influence to protect society and environment globally?

Partner Integration

Cradle-to-Cradle Principle

Knowledge Management & Co-creation Capacity

Customer Collaboration

Global Stewardship

Aiming at realizing the regenerative world through principle-led and responsible behaviour and collaboration across the supply chain ecosystem

Fig. 4.2 The circular supply chain paradigm

1. The principle of sustainability has to be made a priority for the CEO and the top management level. It needs to be rooted in the leadership values.
2. The circular supply and value chain has to be viewed as both the fundamental goal and the business strategy.
3. A framework for decision-making has to be established that illustrates how trade-offs between business ambition and societal and environmental concerns can be addressed.
4. Operating principles have to be agreed among all parties along the CSC, including collective actions in crisis situations.
5. Models for partner integration and customer collaboration enabling cocreation along the supply chain have to be defined. They serve as reference for all joint activities.
6. Platforms for communication and knowledge transfer have to be established.
7. Performance matrixes to measure compliance with the circular principles and business objectives need to be set up and agreed between parties.
8. Governance for implementation and development of the circular business approach must be prepared and agreed.

Establishing the CSC requires a holistic, well-conceived, and thought-through approach, from values and vision to strategies, principles, frameworks and operating models, platforms and performance matrixes, and governance mechanisms. All stakeholders need to commit to the circular concept and support its realization. The vision and the goal of the CE&S have to become like a second skin for everyone and have to be accepted not only as an important factor for success, but as a necessity for ensuring our survival.

Literature

A greener refill. (2016). *Environmental benefits: Reuse & recycling ink and toner cartridges.* Accessed February 23, 2016, from http://www.agreenerrefill.com/the-benefits-of-recycling

Arunarwati, M., Belina, P., Peter V., Turubanova, S., Stolle, F., & Hansen, M. C. (2014, June 29). Primary forest cover loss in Indonesia over 2000-2012. *Nature.* Available at http://www.nature.com/articles/nclimate2277.epdf

Ashton, H. (2014, December 17). The rise of product-as-a-service in manufacturing and some of the technology impacts. *IDC Community.* Accessed September 15, 2016, from https://idc-community.com/manufacturing/manufacturing-value-chain/the_rise_of_product_as_a_service_in_manufacturing_and_some_of_the_te

Audi Encounter. (2014). Thankful outcome. http://audi-encounter.com/magazine/technology/01-2014/102-tankful-outcome; Joule Unlimited. Joule + Audi. Accessed February 22, 2016, from http://www.jouleunlimited.com/joule-audi

Beck, F. (2013). Ein Vorreiter kommt nicht auf Trab. *Verkehrsrundschau, 48,* 26 ff.

Braw, E. (2014, May 5). Free lunch, anyone? Foodsharing sites and apps stop leftovers going to waste. *The Guardian.* Accessed March 04, 2016, from http://www.theguardian.com/sustainable-business/free-food-sharing-leftovers-surplus-local-popular

Brooks Blog. (2015, April 19). *Recycle, reuse and donate your running shoes this earth day.* Accessed February 22, 2016, from http://talk.brooksrunning.com/blog/2013/04/19/recycle-reuse-and-donate-your-brooks-running-shoes-this-earth-day/

Cerasis. (2015). *Reverse logistics and cerasis* (Cerasis Whitepaper, p. 3)

Chapman, M. (2016, April 18). *O'Neill reveal the Blue Collection, a range of clothing made from recycled beach plastic* (Surf-Report). Accessed April 25, 2016, from https://surf-report.co.uk/oneill-reveal-the-blue-collection-a-range-of-clothing-made-from-recycled-beach-plastic-4133/

Chrisafis, A. (2016, February 4). French laws forbids food waste by supermarkets. *The Guardian.* Accessed March 04, 2016, from http://www.theguardian.com/world/2016/feb/04/french-law-forbids-food-waste-by-supermarkets

Cima, R. (2016). How much recycling actually gets recycled: Nitty-gritty. *Alumni Magazine,* Stanford University. Accessed February 23, 2016, from https://alumni.stanford.edu/get/page/magazine/article/?article_id=47701

Clancy, H. (2015, June 2). Meet the startup keeping big-box returns out of landfills. *GreenBiz.* Accessed February 23, 2016, from https://www.greenbiz.com/article/meet-startup-keeping-amazon-and-ebay-returns-out-landfills

Ellen MacArthur Foundation. (2016). *The circular economy 100.* Available at http://www.ellenmacarthurfoundation.org/ce100/the-programme/enabling-collaboration

Endomines. (2016). Accessed February 23, 2016, from http://www.endomines.com/index.php/from-discovery-to-mine

Environmental Protection Agency. (2015). Newsroom, News Releases. "Trash and recycling, EPA report shows progress on E-recycling and identifies opportunity to advance G7's recognition of circular economy". Release date: 06/17/2015, to be found at http://yosemite.epa.gov/opa/admpress.nsf/Press Releases - Trash and Recycling

Environmental Protection Agency. (2016a). *Electronics donation and recycling.* Accessed February 24, 2016, from http://www.epa.gov/recycle/electronics-donation-and-recycling

Environmental Protection Agency. (2016b). *Municipal solid waste.* Accessed February 18, 2016, from https://archive.epa.gov/epawaste/nonhaz/municipal/web/html/index.html

Erviani Komang, N. (2012, December 12). US $10 million investment in plastic recycling factory in Bali. *The Jakarta Post.* Accessed March 07, 2016, from http://www.thejakartapost.com/news/2012/12/12/us10-million-investment-plastic-recycling-factory-bali.html

ETA. (2016). *Transportation and air quality, transportation and climate, regulations and standards, regulations and standards: Heavy-duty.* Accessed February 25, 2016, from https://www3.epa.gov/otaq/climate/regs-heavy-duty.htm

Ethical Trading Initiative. (2016). The base code can be found here http://www.ethicaltrade.org/eti-base-code

European Commission. (2016). *Environment, circular economy.* Accessed January 12, 2016, from http://ec.europa.eu/environment/circular-economy/index_en.htm

European Commission, Climate Action. (2016a). *Reducing emissions from transport.* European Commission. Accessed September 16, 2016, from http://ec.europa.eu/clima/policies/transport/index_en.htm

European Commission, Climate Action. (2016b). *The EU emission trading system.* Accessed April 14, 2016, from http://ec.europa.eu/clima/policies/ets/index_en.htm

Fair Trade USA, Media Kit. (2016). *Understanding fair trade* (p. 2).

Fairminded. (2016). Accessed February 23, 2016, from http://www.fairmined.org/what-is-fairmined/

Future Aerials. (2016). www.futureaerial.com

Gillis, J. (2016, September 3). Flooding of coasts, caused by global warming, has already begun. *New York Times.* Accessed September 16, 2016, from http://www.nytimes.com/2016/09/04/science/flooding-of-coast-caused-by-global-warming-has-already-begun.html?emc=edit_ta_20160903&nlid=62963880&ref=cta&_r=1

Gosden, E. (2016, February 24). Ikea plans mushroom based packaging as eco friendly replacement for polystyrene. *The Telegraph.* Accessed March 06, 2016, from http://www.telegraph.co.uk/news/earth/businessandecology/recycling/12172439/Ikea-plans-mushroom-based-packaging-as-eco-friendly-replacement-for-polystyrene.html

Grunder, D. (2012, August). *Wasted: How America is losing up to 40 percent of its food from farm to fork to landfill* (p. 1). Natural Resources Defence Council (NRDC) Issue paper

Hawks, K. (2006, Winter/Spring). What is reverse logistics? *The Reverse Logistics Magazine,* p. 12

ILO. (2016). http://www.ilo.org/global/about-the-ilo/lang--en/index.htm

International Council on Mining & Metals [ICMM], Communities and Small-Scale Mining (CASM), International Finance Corporation's Oil and Gas and Mining Sustainable Community Development Fund (IFC CommDev). (2009). *Working together: How large-scale miners can engage with artisanal and small-scale miners* (p. 9). Available at https://www.icmm.com/document/789

Invest in China. (2016). Available at http://www.fdi.gov.cn/1800000121_39_597_0_7.html

Lehmacher, W. (2013). *Wie Logistik unser Leben prägt* (p. 63 ff). Springer Gabler

Lexmark. (2016). http://csr.lexmark.com/supplies.html

Lyons Hardcastle, J. (2015, April 20). Apple buys forest to produce sustainable packaging. *Environmental Leader.* Accessed April 29, 2016, from http://www.environmentalleader.com/2015/04/20/apple-buys-forests-to-produce-sustainable-packaging/

Marchi, A., & Parekh, E.-J. (2015, December). How the sharing economy can make its case. *McKinsey Quarterly.* Accessed January 04, 2016, from http://www.mckinsey.com/insights/strategy/how_the_sharing_economy_can_make_its_case

Marks & Spencer. (2016). *Waste & circular economy.* Accessed January 12, 2016, from http://corporate.marksandspencer.com/plan-a/our-approach/business-wide/waste-and-circular-economy

Mathews, J. A., & Tan, H. (2016, March 23). Circular economy: Lessons from China. *Nature.* Accessed April 07, 2016, from http://www.nature.com/news/circular-economy-lessons-from-china-1.19593

Ministère de l'Environnement, de l'Energie et de la Mer. (2016). Available at http://www. developpement-durable.gouv.fr/IMG/pdf/Information_CO2_ENG_Web.pdf

Prakash, A. (2016, February 16). Uber may deliver latest blow to European parcel firms. *Reuters.* Accessed March 06, 2016, from http://www.reuters.com/article/us-europe-markets-parcel-idUSKCN0VP1MB

Reuter, B., Kutter, S., Kempkens, W., & Busch, A. (2013, July 31). *Treibstoffe—Müll tanken und los.* Handelsblatt/WirtschaftsWoche Online. Accessed February 22, 2016, from http://www. handelsblatt.com/technik/das-technologie-update/energie/treibstoffe-muell-tanken-und-los/ 8574054.html

Reuters. (2016, January 21). *China to allocate $4.6 bln to shut 4,300 coal mines—Xinhua.* Accessed February 29, 2016, from http://uk.reuters.com/article/china-coal-closures-idUKL3N1551R4

Rice, P. (2012, January 11). *Fair trade USA: Why we parted ways with Fair Trade International.* tripplePundit—people, planet, profit. Accessed February 22, 2016, from http://www. triplepundit.com/2012/01/fair-trade-all-fair-trade-usa-plans-double-impact-2015/

Ro, S. (2013, April 24). *Here's how many tons of rock you have to mine just for an ounce of gold.* Businessinsider.com. Accessed February 23, 2016, from http://www.businessinsider.com/tons-of-rock-for-an-ounce-of-gold-2013-4

Rolls Royce. (2016). *Metal recycling.* Accessed March 20, 2016, from http://www.rolls-royce. com/sustainability/performance/case-studies/metal-recycling.aspx

Salmon, J. (2015, September 30). *Telematics—Insights into consumer expectations can aid insurers' compliance with financial services regime.* Out-Law.com. Accessed February 22, 2016, from http://www.out-law.com/en/articles/2015/september/telematics--insights-into-consumer-expectations-can-aid-insurers-compliance-with-financial-services-regime

Shearman, S. (2013, April 02). Consumers getting more environmentally conscious. *PR Week.* Accessed March 07, 2016, from http://www.prweek.com/article/1276219/consumers-getting-environmentally-conscious

Smithers, R. (2014, September 3). Global fairtrade sales reach £4.4bn following 15 percent growth during 2013. *The Guardian.* Accessed February 22, 2016, from http://www.theguardian.com/ global-development/2014/sep/03/global-fair-trade-sales-reach-4-billion-following-15-per-cent-growth-2013

Stackpole, B. (2015, May). *IoT-enabled product as a service could transform manufacturing.* TechTarget.com. Accessed March 04, 2016, from http://internetofthingsagenda.techtarget. com/feature/IoT-enabled-product-as-a-service-could-transform-manufacturing

Sweden.se. (2016). *The Swedish recycling revolution.* Accessed February 22, 2016, from https:// sweden.se/nature/the-swedish-recycling-revolution

Tan, M. (2015, May 20). Asphalt mix made with recycled printer toner paves way for eco-friendly roads. *The Guardian.* Accessed February 23, 2016, from http://www.theguardian.com/ australia-news/2015/may/20/asphalt-mix-made-with-recycled-printer-toner-paves-way-for-eco-friendly-roads

The Boston Consulting Group. (2015, March). *The ups and downs of gold recycling* (p. 4/5)

The Economist. (2013, March 7). *Peer-to-peer rental: The rise of the sharing economy.* Accessed September 15, 2016, from http://www.economist.com/news/leaders/21573104-internet-every thing-hire-rise-sharing-economy

The Economist. (2015). *The cost of inaction.* Available at http://www.economistinsights.com/ sites/default/files/The%20cost%20of%20inaction.pdf

Think Progress. (2016, May 9). Why used electric car batteries could be crucial to a clean energy future. Accessed September 15, 2015, from https://thinkprogress.org/why-used-electric-car-batteries-could-be-crucial-to-a-clean-energy-future-6ab9a2308cdb#.4hkzus9do

Triple Pundit—people, planet, profit. (2015, July 21). *Levi's drives the circular economy for clothing.* Accessed January 12 and April 20, 2016, from http://www.triplepundit.com/podium/ levis-circular-economy/

U.S. Department of Justice, Office of the Inspector General. (2010, October). A review of Federal Prison Industries' electronic-waste recycling program

U.S. Securities and Exchange Commission. (2016). https://www.sec.gov/spotlight/dodd-frank/speccorpdisclosure.shtml

Unilever. (2016). *Unilever sustainable living plan, Sustainable sourcing*. Accessed February 22, 2016, from https://www.unilever.com/sustainable-living/the-sustainable-living-plan/reducing-environmental-impact/sustainable-sourcing/

United States Environmental Protection Agency. (2016). *Municipal solid waste*. Accessed January 22, 2016, from https://archive.epa.gov/epawaste/nonhaz/municipal/web/html/index.html

Verkehrsrundschau #51-52/2014. *Wir arbeiten in den Prozessen rückwärts* (p. 20 ff)

Volkswagen AG. (2016). *The life cycle of a Car—Environmental commendations document progress.*

Walk Free Foundation. (2014). *Global Slavery Index 2014*. Accessed April 29, 2016, from http://www.globalslaveryindex.org/findings/

Watch Renault's video on YouTube. (2015). https://www.youtube.com/watch?v=4zSWTT0qCzQ

Wong, K. (2015, December 14). Entrepreneurs turn billion dollar seafood waste into profitable products. *The Guardian*. Accessed March 10, 2016, from http://www.theguardian.com/sustainable-business/2015/dec/14/us-fishermen-turn-billion-dollar-seafood-waste-into-profitable-products

World Economic Forum. (2015, January 24). *Beyond supply chains—Empowering responsible value chains*. Available at https://www.weforum.org/reports/beyond-supply-chains-empowering-responsible-value-chains/

Shaping the Future of Supply Chains

5

Smart supply chains are agile supply chains. Establishing agile supply chains requires vision and solid change management capabilities. Speed, flexibility, data management, and collaboration are key characteristics of modern supply chains.

5.1 Agile Supply Chains

Agile supply chains are characterized by end-to-end design, high visibility and transparency, alignment, and close collaboration – from design to planning to execution. Agile supply chains act and adapt quickly; they affect performance and cost. Companies with agile supply chains feature low costs of transport, handling, procurement in production, or supply of services. They also manage an optimal allocation of inventory across distribution networks (Dubeauclard et al. 2015).

The German Association of Logistics (*Bundesvereinigung Logistik*, BVL) has outlined and described nine focus areas critical for building agile supply chains:

1. *Transparency*: Only by monitoring how customer expectations, market conditions, and technology are changing can we analyze the potential impact early on and adapt strategies when necessary.
2. *Conscious make-or-buy decisions*: Critical competencies and services should stay with the company. Where this is not possible, at least two suppliers or service providers should be available.
3. *Selection of suppliers*: Suppliers' abilities to perform and collaborate defines the efficiency and robustness of the supply chain. In critical situations, a company has to rely on their partners. The right choice is therefore key.
4. *Joint development*: Design plays an important role in establishing the supply chain. Therefore, collaboration is a decisive success factor. The sooner the service provider and supplier are involved in design considerations of new products and processes, the better the alignment of all players within the supply

© Springer International Publishing AG 2017
W. Lehmacher, *The Global Supply Chain*, Management for Professionals,
DOI 10.1007/978-3-319-51115-3_5

chain ecosystem. When players are attuned to one another, they can initiate optimization strategies or quickly counteract critical situations. Competitive advantages can also be identified and strengthened.

5. *Independent development units*: Involvement of suppliers and service providers in the design of services and products can be structured through a legally independent development entity. Due to its independence, processes can be considerably sped up and allow for new creative approaches in product and process design.

6. *Talent pools and competence networks*: Establishing an overall human resources (HR) operations across all locations will support knowledge management and avoid bottlenecks. HR pools provide flexibility to assign and engage employees where needed based on their specific skills and willingness to step in.

7. *Strategic supplier management*: The close collaboration with suppliers in planning processes helps to enhance the quality of products and services as well as create capacities and flexibility.

8. *Dynamic quality assurance*: If players are accountable for their activities and decisions, quality will increase; self-organizing work structures will raise individual accountability and collective responsibility across teams.

9. *Dynamic, self-regulating project groups*: By simply fixing a goal, members of a project group will create a roadmap themselves. As today's dynamic world requires quick action, the advantage is clear: teams more flexible in their decisions can react faster to changing frameworks, situations and requirements (Bundesvereinigung Logistik 2012).

Designing and managing agile supply chains requires knowledge. Global logistics company, *DB Schenker*, fosters knowledge-sharing and knowledge-transfer by coaching, thematic partnerships and cooperation, and an open feedback culture. An internal platform features an encyclopedia, accessible to all logged-in users. They can publish, update, or research articles, reports, papers, user manuals for IT tools, and newspaper articles. The platform also contains an interactive membership directory, a workflow-based idea management, lists for property, office supply, and private bids (DB Schenker 2013). This enables the employees to interact fast, within and beyond their own scope of work and responsibility.

5.2   Big Data and Data Management

Agile supply chains cannot be realized without advanced data management. Data analysts at IDC predict that, from 2005 to 2020, the digital universe will have grown by a factor of 300, from 130 exabytes to 40,000 exabytes (40 trillion gigabytes, 40 zettabytes) (IDC 2012). The data will come from all different sources, ranging from smartphones to email to home gas and electricity meters. In the near future, the amount of data communicated between vehicles and from smart vehicles to urban and national infrastructure will increase as well.

In Germany alone, DHL delivered about 3.4 million parcels every day in 2014. They record up to 150 parameters for each one, such as size, weight, destination, or contents, which amounts to over 300 million pieces of new information every day (Kückelhaus 2014). Therefore, logistics service providers will become vital information suppliers. Modern logistics equals e-logistics. Without data, the management of parcels of this magnitude is just not manageable anymore. The same is true for nearly any activity regarding the supply chain. Logistics companies and any player upstream and downstream in the supply chain collect, analyze, and store data not just for their own purpose. DHL and competitors distribute shipment data to customers, thereby contributing to company-wide supply chain visibility (SCV) and more efficient planning and management of the supply chain as a whole. Data has been around for many years, of course. However, only today are we able to analyze it in massive amounts and use it to back up conclusions that form the basis of decision-making and action.

Transport and logistics companies, as well as other supply chain service providers, can mitigate risk and detect fraud, plan staffing and capacity, and much more. UPS, for example, gathers data on packages, pieces, and parts. Huge amounts of data are monitored constantly to assess performance and to redesign drivers' routes, an initiative that is called On-Road Integration Optimization and Navigation (ORION). It relies on data from online maps and reconfigures a driver's pickups and drop-offs in real time. Up to now, the project has led to savings of more than 8.4 million gallons of fuel by cutting 85 million miles off daily routes. UPS estimates that saving only one daily mile per driver saves the company $30 million (SAS 2016). This is a great example of how big data and smart technology can impact the triple bottom line.

5.3 The Autonomous Supply Chain

The smartest and most agile supply chain is autonomous. Smart means fluid, clean, safe, secure, and cost-efficient. Autonomous supply chains can deliver on all of these objectives. The autonomous supply chain needs to take all data into account to navigate through a highly complex world by continuously making correct decisions. In addition, the autonomous supply chain must learn, including taking the effects of past decisions into future decision-making processes.

Indoor, automated guided transport (AGT) systems and automated guided vehicles (AGVs) have been around since the early 1950s, when automated systems optimized material flows and helped to reduce cost in manufacturing and warehousing in the USA. Today, AGT systems are widely used in the supply chain. *Amazon Robotics* (formerly Kiva), for example, builds systems that increase picking efficiency in warehouses through robots transporting the entire shelf to a picker, which allows the person to stay in one place. Autonomous loading and transport equipment also helps to reduce cost and errors in the supply chain. Outdoors, AGVs move containers around ports to shorten routes and reduce trips made with empty vehicles. For example, the Port of Los Angeles underwent a major improvement program and now calls *TraPac*, the most technologically advanced container terminal on the West Coast of the USA, its home.

Beyond private company grounds, automated transport has been successfully tested on public roads. Six convoys of platoons, in groups of two or three trucks, communicating wirelessly and driving close behind one another, arrived in the port city of Rotterdam in April 2016 (Supply Chain Digest 2016). The Ministry of Transport in Singapore and the PSA Corporation, formerly known as Port of Singapore Authority, seek proposals to develop an autonomous truck platooning system (Singapore Government 2015). In 2013, inventor and tech entrepreneur Elon Musk envisioned the Hyperloop vacuum tube, previously discussed, to move passengers and cargo in 35 min from San Francisco to Los Angeles. As an alternative to surface transportation on roads or through tubes, experts in Switzerland have presented a plan to move goods across the country in a gigantic underground tunnel (Swissinfo.ch 2016). On the waters, the cargo ships of the future are expected to be crewless and remote-controlled (Hutt 2016).

The so-called last mile, the final delivery to businesses and consumers, is probably the most complex segment of the supply chain, particularly in diverse urban environments, with cars and bicycles and children playing on the streets. Some have proposed to address this environment with drones, but the start-up company *Dispatch* has developed a robotic last mile solution on wheels, which can move a total of 100 pounds (Dickey 2016). In order to learn to move safely alongside other vehicles and citizens, the robot is equipped with sensors and artificial intelligence technology. Intelligent lockers and boxes placed at or near the receiver's and sender's locations complement the autonomous delivery and pickup system.

In parallel, companies like *Immense Simulations* are working on building the brain and software for the autonomous world, able to coordinate thousands of autonomous units on, under, and above the ground, on the rivers, lakes, and oceans, moving passengers and cargo (Olson 2016). The autonomous end-to-end supply chain is almost complete. Autonomously, raw materials are extracted in automated mines,[1] then transported to smart factories,[2] and from there brought by truck platoons, or long-distance drones or through tunnels, to purchasers or the automated distribution centers of wholesalers, retailers, or e-commerce companies. The business customers and consumers receive the orders via drones, robots, or urban tubes delivered into offices, homes, or smart boxes – or soon, via private self-driving cars sent for pickups to one or more sorting centers.

The autonomous movement commenced in the early 1950s and is now in full swing. Seventy percent of 1433 consumers surveyed in the USA think they will order their first drone-delivered package within the next 5 years (Walker Sands

---

[1]Further information can be found at http://www.cat.com/en_US/support/operations/technology/cat-minestar/minestar-in-action/auto-haul-imp-mine-site-safety.html

[2]Industry 4.0: the current trend of automation and data exchange in manufacturing technologies (Source: Wikipedia)

Communication 2016). Eighty-eight percent of policymakers expect autonomous vehicles to gradually become reality within the next 10 years, based on a recent global self-driving vehicle study by the World Economic Forum in collaboration with Boston Consulting Group. Many of the policymakers interviewed consider goods delivery as one of the key applications for autonomous transportation in their city. Sixty percent of the policymakers in the study expect a ban for private cars in a significant portion of their cities over the next 15 years (World Economic Forum and Boston Consulting Group 2015). Will this be limited to private vehicles? Probably not. Over time cities will further regulate goods deliveries. One pain point for citizens today is daytime deliveries which require double parking and add to the city's overall congestion. Therefore, not only transportation and logistics companies, but shippers need to prepare for the autonomous future.

The global self-driving vehicle study yielded positive news in respect to the environment and our health: 66% of the 5500 consumers surveyed stated that, in their minds, self-driving cars should be electric or hybrid. In light of all the benefits and challenges of the eventual fully autonomous supply chain, it is important to understand the full implication on the mobility system as a whole. Building a pilot city somewhere in the world with fully fledged autonomous goods and personal mobility would represent a major step forward towards cleaner, safer, and more efficient cities.

5.4 Opportunities Along a New Path to Progress

The global supply chain is an ecosystem supporting the core of the modern economy and society. It delivers the goods we need every day. The design and management of this ecosystem requires not only specific knowledge and skills, commonly known as procurement, sourcing, transport, and logistics, but also good global stewardship to secure the quality of life for customers and citizens.

Strong leadership by CSCOs is necessary to steer the ecosystem in the right direction. As more options for design, planning, and management become available through technology and a growing wealth of connections and information flows through society, CSCOs need to convert new knowledge into new models. The field of supply chain is no longer just about technical or operational processes, but is rather a renewed set of core principles around the extraction, movement, production, and repurposing of raw materials, intermediates, parts, and finished products.

Embracing a new set of principles in global supply chains can lead society to new pools of opportunities. The Ellen MacArthur Foundation has identified "a $1 trillion opportunity for businesses worldwide in adopting a circular economy." According to the foundation, in the future, we will be able to use web-based tools to measure and document a set of indicators, such as the Material Circularity Indicator (MCI) that can speak of the circularity of products at multiple levels. Hence, all materials and goods will be traceable in the future, and those without a transparent "genetic code" will increasingly face scrutiny that will impede their fluid travel through the ecosystem. What products are made of, their use cycles, and the subsequent repurposing will be as transparent as their source of origin. We will be able to see where commodities came from, how a good was made, and how much carbon dioxide went into the atmosphere from producing and shipping the object.

A system like this will constantly provide data on materials, on products and processes, and on usage, maintenance, repair, reuse, and recycling. As a result, a measurable and more meaningful way of making and using things will surface, with unlimited opportunities for new business models. Supply chains can become regenerative, self-learning ecosystems that will support the most suitable and balanced actions in the best interest of all stakeholders involved.

5.5   Challenges Limiting Ecosystem Leadership

The world has experienced decades of advancing global integration. Increasingly open borders and trade and investment partnerships have strongly contributed to the prosperity and wealth of people and nations. International organizations and agencies have not only supported global growth but also established institutions in charge of dealing with the risks of reducing national barriers. Organizations have developed international ties and many platforms of collaboration to fight crime and terror.

Most recently, though, large parts of the world have gone through a phase of increasing global disintegration. In light of massive flows of migrants seeking asylum in Europe, 65 countries have erected fences on their borders, which is "four times as many as when the Berlin Wall was toppled" according to *The Daily Mail* (Tomlinson 2015). In December 2015, the BBC wrote: "EU border security becomes new mantra" (Peter 2015). The Brexit referendum, discussions about the exclusion of Greece from the Eurozone and the beginning of the construction of fences along the green borders of barrier-free Schengen are events that are challenging the capacity of system leaders to scale up the dialogue for regenerative supply chains.[3]

On June 14, 1985, the picturesque town of Schengen in Luxembourg welcomed five European countries for the signature of the agreement which led to the creation of Europe's borderless area. Thirty years later, in September 2015, Hungary has blocked migrants from onward travel to the rest of Europe and constructed a 4 m tall fence along sections of the border with Serbia, a country which is not part of the Schengen area. In April 2016, Austria has begun building an anti-migrant barrier across the Brenner Pass at the Italian border, putting an end to hope on one side and maybe reducing fears on the other. Beyond Europe, increasing fear of terrorists in the USA has led to the reintroduction of a visa for "certain Europeans," meaning those who have traveled to the Middle East or have dual citizenship (New Europe 2016).

---

[3]The Schengen area is the area including 26 European countries that have abolished passport and any other type of border control at their mutual borders. (Source: Wikipedia)

What is the expected impact of reestablishing these barriers? Citizens will face long-forgotten burdens: northern Europeans, for example, will experience long traffic jams at the Brenner Pass on the way to their holiday destinations in the South. Labor markets will also be affected: 1.7 million people cross European borders every day to get to work (Wang 2016). Consumer prices would rise due to the forced slowdown and necessary adjustments along the supply chain. Waiting and inspection times at the borders would need to be factored into the prices of goods. This is particularly true for the highly cost-optimized just-in-time concepts largely applied in global manufacturing in the automotive industry. Many of the products made available by bilateral and multilateral agreements risk disappearing from supermarket shelves.

Disintegration could also affect the competitive position of nations. European countries might find themselves in a disadvantaged situation given that Asia is continuing to integrate. What if TPP arrives and Schengen leaves? Overall, there could be explosive geopolitical risks involved – with Crimea, Ukraine, and the new Chinese islands in the South China Sea heating up the debate. As new fences go up across Europe, what tensions will result in countries such as Spain, Italy, and Greece, who are being left more or less alone with the new waves of migrants coming across the Mediterranean Sea?

The flows of refugees have many effects. The International Chamber of Shipping (ICS) pointed out that between January 2014 and September 2015, more than 1000 merchant ships have helped rescue more than 65,000 people. That's more than one in ten of the estimated 585,000 migrants and refugees who crossed the Mediterranean during that period (Saul 2015). At the end of 2014, the ICS updated their "Guidance on Large Scale Migrant Rescue." It states, however, that "the revision of the Guidelines does not in any way suggest that shipping companies or their crews are reconciled to the continuing failure of governments to provide adequate state-backed rescue resources, as required by international law" (International Chamber of Shipping 2016). Often, seamen have to cope with inadequate accommodation, catering, and medical services for, and the deaths of, rescued people.

The German Shipowners' Association (VDR) therefore calls for the expansion of the geographic area covered by Operation Triton in the Central Mediterranean, where *Frontex*, an agency of the European Union established in 2004 to manage the cooperation between national border guards securing its external borders, operates three open sea patrol vessels, two coastal patrol vessels, two coastal patrol boats, two aircraft, and one helicopter. The benefits of an expanded theater of operation mean that Frontex vessel operators would be able to transfer refugees faster to either medical support teams or transport vessels (German Shipowners' Association (VDR) 2016).

The sheer volume of refugees is impacting European logistics services. In 2015, truck drivers had to deal with refugees who tried to cross the North Sea through the Eurotunnel from Calais in France to the UK. In order to get on the trucks, refugees broke seals or locks and threw out cargo. Authorities can prosecute drivers when illegal immigrants are discovered in the trailers. Drivers and/or company owners have to prove that they are not involved in human trafficking. Drivers caught with illegal stowaways on board their vehicles are currently fined up to £2000 ($2.825) per person discovered. According to the British Freight Transport Association (FTA), around 40 concealed migrants a day were estimated to have entered Dover in the summer of 2015. The tunnel was closed time and again because people kept trying to enter the passage. Long traffic holdups on both sides of the channel were the result. The FTA estimates that industrial action and increased migrant activity throughout the summer of 2015 cost the logistics industry an estimated £21 million ($29.67 million) for delays, canceled sailings, and interrupted tunnel crossings (Freight Transport Association 2016).

But the idea of closing borders to improve security may be illusive at best. How safe has the world been with more barriers? Did borders protect Italy from the onslaught in the 1970s of the Red Brigades, Spain from the ETA, Germany from the Red Army, and France from GIA? Did borders protect the USA from attack on 9/11? How effective were the high metal fences and walls, barbed wire, alarms, anti-vehicle ditches, watchtowers, automatic booby traps, and minefields along the inner German border from 1945 to 1990? The threat often lies within: "Not one Paris attacker has been identified as a Syrian refugee," Megan Specia wrote on *Mashable* (Specia 2015).

Social media helps terrorists organize themselves and recruit new fighters (Burke 2016). The FBI uses Internet surveillance software like *Carnivore* to identify and stop attackers. Organizations such as the Search for International Terrorist Entities scan propaganda material and training manuals and share insights with other organizations. Technology trumps technology. The Internet has the potential to erase borders while reducing risks. The more people are active on the net, the better economic value can be extracted and (potential) terrorist activities monitored, which does not come without its own concerns and complexities, as the discussion between Apple and the FBI over the access to the iPhone used by one of the shooters in the San Bernardino attacks in 2015 shows (Kharpal 2016).

Interpol, the International Criminal Police Organization, has strong links with Europol, the organization coordinating the local police forces across Europe. Within countries, ministries and agencies are increasingly working together. Germany, for example, has established the Joint Counterterrorism Center (GTAZ), an autonomous authority and cooperation platform used by 40 internal security agencies.[4] The private sector has also launched initiatives to protect staff and assets against terrorism and other threats across the globe. Since the attacks of 9/11, security measures have been tightened. Today, individuals and companies are checked against the sanction lists of the USA and Europe. Employees appearing on the lists are no longer allowed to be paid a salary, and companies are excluded from doing business. Though, as the Panama papers show, we have not yet closed all the back doors.

---

[4]Further information can be acquired at https://www.verfassungsschutz.de/en/fields-of-work/islamism-and-islamist-terrorism/gtaz-en

Governments have the obligation to protect citizens and the right to control borders. However, what are the effects of the potential disintegration on citizens, migrants, and the economy? The Bertelsmann Foundation warns that reestablishing permanent border controls in Europe could produce losses of up to 1.4 trillion euros over 10 years (Charter 2016). Political leaders and society need to understand and be mindful of the impact of our decisions on the economy and finally on the society. With increasing fears and instability, blurring borders, and rising volatility, governments and the business community need to deal with the undesired reactions, like rejection of technology and antiglobalization movements.

5.6   Coming Back to Prosperity and Peace

Globalization and the global supply chain can be a guarantee for dignified living conditions for all human beings on planet earth. Both are key enablers for societal and economic progress. The avoidance of risk, the fight against criminal activity, and terrorism can be supported by new technology. Visibility along the supply chain will make bad behavior, corruption, terrorist attacks, and other criminal actions increasingly difficult. Powerful tools to analyze big data will help to identify trends, patterns, and threats. Supply chain managers and governmental agencies will be able to prepare fast and act proactively, with some systems even acting autonomously.

Governments and businesses need to implement new solutions that will include future generations and those who were left behind by the current path of development. The efforts of the Canadian government, for example, to improve the position of the middle class show that this might not be that easy, "either because other countries' workers will still work for less, or because the sources of rising living standards have been exhausted, or because robots are taking over menial (and not-so-menial) jobs, or simply because the jar of magic pixie dust that defined the last 70 years of life in North America is finally empty. And if the dream of a constantly rising living standard for the middle class no longer holds, then a government's job is to search for more creative ways to make whatever new reality emerges less arbitrary, or at least less unpleasant" as Christopher Flavelle wrote in *Bloomberg* magazine (Flavelle 2016).

The business world has become flat, because Internet platforms are flat. However, when the shipping of physical goods comes into play, crossing oceans and mountains is very real. And in navigating and shaping bureaucracy, the capacity to influence the status quo and the future of the earth is very clear.

The supply chain ecosystem can be regenerative in nature, capable of renewing itself with the circular supply chain (CSC) at its core. More a network than a chain, the supply chain has multiple functions and effects in society. Thanks to the Internet, the supply chain connects a universe of possibilities, containing millions of suppliers and billions of customers, consumers, and prosumers, which, just like stars, shine bright, flicker for a short period of time, and eventually disappear. They are all connected through the web and face various disruptive factors.

In a circular world, chief supply chain officers (CSCOs) need to orchestrate suppliers, manufacturers, vendors, consumers, and recyclers. They will need to design and plan systems that can foresee and nimbly buffer potential disruptions and shocks and minimize costs and environmental burden. At the same time, the system must maximize the sustainable procurement, sourcing, handling, transporting, manufacturing, and repurposing. Only in this way can they ensure the good of business customers, consumers, citizens, society, and the planet and secure long-term revenue and profits for the company and returns for investors.

Consumers will become still more empowered and more influential and must use their power to improve the system too. Every player in the supply chain ecosystem will be able to learn about almost anything related to a product, from the origin of materials to the style of manufacturing and transport. The blockchain will enable a new level of transparency in the supply chain to reduce fraud and corruption and raise compliance and confidence. This might give new air to global trade agreements. Through social media or modern collaboration systems, all stakeholders are connected, can communicate, and show likes or dislikes. Cocreation with consumers will enlarge the diversity of possibilities and solutions. Innovation and transaction will grow on all three levels: human-to-human cocreation, human-to-machine, and machine-to-machine interaction. Artificial intelligence (AI) and robots with deep-learning capabilities will unlock potential as yet unthought-of.

We need to progress responsibly. Protectionism, the growing global population, scarcity of resources, and environmental pollution require a new way of thinking and acting, with the circular supply chain ecosystem at its core. Unequal distribution, social tensions, geopolitical conflicts, and wars cause a backlash against globalization and modern civilization. Environmentalists and antiglobalization groups aren't all wrong: there is plenty of room for improvement. A new mindset and course of action to make our current economic and societal system circular are needed to prevent these tensions from worsening. More of the financial gains of globalism should be put towards supporting and retraining those who lose out when jobs move, for example.

But, even without improvement, the effects of globalization and global connectivity have been positive overall. The increase in living standards over the past century is undeniable, not only in emerging markets, but in Western Europe and the USA too, caused largely by the exponential expansion of cross-border trade after the World War II. Foreign investment delivers knowledge, technology jobs, and better access to goods in both directions, while consumers, and often those worst-off, pay the price when tariffs on foreign goods are raised. Further, migrants must be acknowledged as net contributors to the GDP of their adoptive countries, which they generally are (The Economist 2016). Technology will help to better ensure security against terrorism and other threats.

The global supply chain must make its contribution without compromise. Forward into the past could be the motto, since until recently, the earth itself was a circular, regenerative system. The future must be circular too. Nature is our mother and our best teacher. Together, we all have to reshape this world. The economy of the future is the circular economy; the society of the future is the circular society.

Literature

Bundesvereinigung Logistik (BVL). (2012). *Supply chain agility—Strategic Anpassung in supply chain management* (p. 8 ff). Bremen

Burke, J. (2016, February 25). *How the changing media is changing terrorism.* The Guardian. Accessed September 16, 2016, from https://www.theguardian.com/world/2016/feb/25/how-changing-media-changing-terrorism

Charter, D. (2016, February 22). *End of Schengen zone 'could cost EU €1.4 trillion over ten years'.* The Times. Accessed September 16, 2016, from http://www.thetimes.co.uk/tto/news/world/europe/article4696489.ece

DB Schenker. (2013, August 2). *Wissensmanagement: Der Mitarbeiter im Mittelpunkt aller Aktivitäten.* Accessed April 07, 2016, from http://www.logistics.dbschenker.de/log-de-de/unternehmen/innovation/Wissensmanagement/Plattform.html

Dickey, M. R. (2016, April 6). *Self-driving delivery vehicle startup dispatch raises $2 million seed round led by Andreessen Horowitz.* Tech Crunch. Accessed September 16, 2016, from https://techcrunch.com/2016/04/06/self-driving-delivery-vehicle-startup-dispatch-raises-2-million-seed-round-led-by-andreessen-horowitz/

Dubeauclard, R., Kubik, K., & Nagali, V. (2015, April). *How agile is your supply chain?* McKinsey Quaterly. Accessed March 14, 2016, from http://www.mckinsey.com/business-functions/operations/our-insights/how-agile-is-your-supply-chain

Flavelle, C. (2016, March 7). *How much can you fight inequality? Canada's about to find out.* Bloomberg. Accessed September 16, 2016, from https://www.bloomberg.com/view/articles/2016-03-07/canada-s-answer-to-trump

Freight Transport Association (FTA). (2016, March 3). *FTA calls for governments to adopt Calais action plan.* Accessed April 06, 2016, from http://www.fta.co.uk/media_and_campaigns/press_releases/2016/20160303-fta-calls-for-governments-to-adopt-Calais-action-plan.html

German Shipowners' Association (VDR). (2016). *Refugee crisis in the Mediterranean Sea.* Accessed April 06, 2016, from http://www.reederverband.de/en/themes-positions/refugees.html

Hutt, R. (2016, July 12). *Remote-controlled and crewless: Is this the cargo ship of the future?* World Economic Forum. Accessed September 16, 2016, from https://www.weforum.org/agenda/2016/07/remote-controlled-and-crewless-is-this-the-future-of-cargo-shipping/

IDC. (2012). *The digital universe in 2010: Big data, bigger digital shadows, and biggest growth in the far east.* Accessed April 07, 2016, from http://www.emc.com/collateral/analyst-reports/idc-the-digital-universe-in-2020.pdf

International Chamber of Shipping. (2016, July 16). *Shipping industry updates guidance on large scale migrant rescue as mediterranean crisis continues.* Accessed April 06, 2016, from http://www.ics-shipping.org/news/press-releases/view-article/2015/07/16/shipping-industry-updates-guidance-on-large-scale-migrant-rescue-as-mediterranean-crisis-continues

Kharpal, A. (2016, March 29). *Apple vs FBI: All you need to know.* CNBC. Accessed September 16, 2016, from http://www.cnbc.com/2016/03/29/apple-vs-fbi-all-you-need-to-know.html

Kückelhaus, M. (2014, February 13). *Drilling deep to find big data riches.* Delivering Tomorrow. Accessed April 07, 2016, from http://www.delivering-tomorrow.com/en/drilling-deep-to-find-big-data-riches/

New Europe. (2016, January 22). *If you have traveled to the Middle East or you have dual citizenship, you need a visa for the U.S.* Accessed September 16, 2016, from https://www.neweurope.eu/article/united-states-reintroduces-visa-for-dual-citizen-europeans/

Olson, P. (2016, February 18). *The powerful brain behind driverless fleets is already being built.* Forbes. Accessed September 16, 2016, from http://www.forbes.com/sites/parmyolson/2016/02/18/this-startup-is-building-a-brain-for-fleets-of-driverless-vehicles/#17fdaab63f62

Peter, L. (2015, December 19). *Migrant crisis: EU border security becomes new mantra.* BBC. Accessed September 16, 2016, from http://www.bbc.co.uk/news/world-europe-35140794

SAS. (2016). *What it is and why it matters.* Accessed March 21, 2016, from http://www.sas.com/en_us/insights/big-data/what-is-big-data.html

Saul, J. (2015, September 21). *Rescue at Sea: In Mediterranean, commercial ships scoop up desperate human cargo.* Reuters. Accessed April 06, 2016, from http://www.reuters.com/investigates/special-report/europe-migrants-ship/

Singapore Government, Ministry of Transport. (2015, October 12). *Joint release by PSA and MOT—Autonomous truck platooning technology to boost port productivity.* Accessed September 16, 2015, from http://www.mot.gov.sg/News-Centre/News/2015/Joint-Release-by-PSA-and-MOT---Autonomous-truck-platooning-technology-to-boost-port-productivity/

Specia, M. (2015, November 23). *Not one Paris attacker has been identified as a Syrian refugee.* Mashable. Accessed September 16, 2015, from http://mashable.com/2015/11/23/paris-attackers-europeans-not-refugees/#bDMzojqqfsqX

Supply Chain Digest. (2016, April 13). *Supply chain news: Successful test of truck platooning in Europe likely to move technology forward, while automatic braking for trucks in the US years away.* Accessed September 16, 2015, from http://www.scdigest.com/ONTARGET/16-04-13-1f.php?cid=10545

Swissinfo.ch. (2016, January 26). *Plans mooted for underground goods tunnel.* Accessed September 16, 2015, from http://www.swissinfo.ch/eng/cargo-tube_plans-revealed-for-under ground-goods-tunnel/41920926

The Economist. (2016, October 31). *Anti-globalists: Why they're wrong.* http://www.economist.com/news/leaders/21707926-globalisations-critics-say-it-benefits-only-elite-fact-less-open-world-would-hurt

Tomlinson, S. (2015, August 21). *World of walls: How 65 countries have erected fences on their borders—Four times as many as when the Berlin Wall was toppled—As governments try to hold back the tide of migrants.* Daily Mail. Accessed September 16, 2015, from http://www.dailymail.co.uk/news/article-3205724/How-65-countries-erected-security-walls-borders.html#ixzz4KOe67cZA

Walker Sands Communication. (2016). *2016 future of retail study.* http://www.walkersands.com/Futureofretail

Wang, C. (2016, March 22). *Brussels attacks: Can Europe keep its open borders?* CNBC. Accessed September 16, 2015, from http://www.cnbc.com/2016/03/22/brussels-attacks-can-europe-keep-its-open-borders.html

World Economic Forum & Boston Consulting Group. (2015, November 24). *Self-driving vehicles in an urban context.* Accessed September 16, 2015, from http://www3.weforum.org/docs/WEF_Press%20release.pdf

Appendix

Books

Lehmacher, W. (2013). *Wie Logistik unser Leben prägt. Der Wertbeitrag logistischer Dienstleistungen für Wirtschaft und Gesellschaft (How logistics shapes our lives)*. Wiesbaden: Springer, Gabler

Lehmacher, W. (2015). *Logistik im Zeichen der Urbanisierung. Versorgung von Stadt und Land im digitalen und mobilen Zeitalter (Logistics in light of urbanisation. Supply of urban and rural areas in the digital age)*. Wiesbaden: Springer, Gabler

Lehmacher, W. (2016). *Steht unsere Versorgung auf dem Spiel? Über terroristische Bedrohungen entlang der Supply Chain (Is our supply at stake? Terrorist threats along the supply chain)*. Wiesbaden: Springer, Gabler

Porter, M. E. (1985). *Competitive advantage: Creating and sustaining superior performance*. New York, NY: Simon and Schuster. Retrieved September 9, 2013

Studies and White Paper

acatech: Menschen und Güter bewegen. (2012). *Integrative Entwicklung von Mobilität und Logistik für mehr Lebensqualität und Wohlstand.*

A.T. Kearney. (2011). *Ideas and insights, China's E-commerce market: The logistics challenges.* Available at https://www.atkearney.com/documents/10192/253176/Chinas_E-Commerce_Market.pdf

A.T. Kearney. (2015). *The 2015 global retail E-Commerce Index.* Available at https://www.atkearney.com/documents/10192/5691153/Global+Retail+E-Commerce+Keeps+On+Clicking.pdf/abe38776-2669-47ba-9387-5d1653e40409

Boston Consulting Group. (2011). *Global aging, how companies can adapt to the new reality.*

Boston Consulting Group. (2015). *The ups and downs of gold recycling.*

Bundesvereinigung Logistik (BVL). (2008). *Zukunft der Logistik-Dienstleistungsbranche in Deutschland 2025.*

© Springer International Publishing AG 2017

W. Lehmacher, *The Global Supply Chain*, Management for Professionals,
DOI 10.1007/978-3-319-51115-3

Bundesvereinigung Logistik (BVL). (2012). *Supply chain agility—Strategic Anpassung in supply chain management.* Bremen

BVL International. (2013). *Trends and strategies in logistics and supply chain management.*

Credit Suisse Research Institute. (2015). *The end of globalization or a more multipolar world?*

Deutsche Post DHL. (Hrsg.). (2016). *Insight on: Risk & resilience.* Bonn

Ernst & Young. (2014). *The road to 2030: A survey of infrastructure development in Russia.* London

Evenett, S. J., & Fritz, J. (2015, November 12). *The tide turns? Trade, protectionism, and slowing global growth.* Centre for Economic Policy Research, CEPR Press. Available at http://www.globaltradealert.org/sites/default/files/GTA18%20The%20Tide%20Turns.pdf

FM Global. (2015). *Resilience Index 2015.*

IBM Global Business Services. (2010, October). *The smarter supply chain of the future.*

Institut der deutschen Wirtschaft. (2015). *Globale Kräfteverschiebung. Kräfteverschiebung in der Weltwirtschaft—Wo steht die deutsche Industrie in der Globalisierung?* Köln

International Energy Agency. (2015). *World energy outlook 2015, executive summary.* Available at https://www.iea.org/Textbase/npsum/WEO2015SUM.pdf

IPSW Strategy Serie. (2015, November). *The New Silk Road—Idea and concept.* Focus on Defense and International Security, Issue No. 390. Wolfgang Lehmacher and Victor Padilla Taylor

KPMG. (2011, May). *On the move in China. The role of transport and logistics in a changing economy.*

McKinsey Consumer & Shopper Insights. (2012, März). *Meet the 2020 Chinese consumer.*

Pew Research Center. (2013, May 29). *Breadwinner Moms.* Available at http://www.pewsocialtrends.org/files/2013/05/Breadwinner_moms_final.pdf

PricewaterhouseCooper. (2011). *Transportation & logistics 2030,* Vol. 4*: Securing the supply chain.* Available at https://www.pwc.ch/user_content/editor/files/publ_trans/pwc_transportation_logistics_2030_vol4.pdf

PricewaterhouseCoopers. (2015, January). *A decade of unprecedented growth—China's impact on the semiconductor industry.* Available at https://www.pwc.com/il/en/assets/pdf-files/china-semicon-2014.pdf

PwC Economics & Policy/Nigeria Economy Watch. (2015, May). *What next for Nigeria's economy? Navigating the rocky road ahead.* Available at https://www.pwc.com/ng/en/assets/pdf/economy-watch-may-2015.pdf

Roland Berger. (2014, August). *Global logistics markets.* Munich

Roland Berger. (2015, May). *The trend compendium 2030.*

Underwriters Laboratories Inc. (2011). *The life cycle of materials in mobile phones.* Available at http://services.ul.com/wp-content/uploads/sites/4/2014/05/ULE_CellPhone_White_Paper_V2.pdf

Wohler Associates: Wohlers Report 2015. ISBN:978-0-9913332-1-9, Fort Collins, Colorado. Available at http://www.wohlersassociates.com

World Bank. (2010). *International LPI global ranking*. http://lpi.worldbank.org/international/global/2010

World Bank. (2014). *Connecting to compete 2014, trade logistics in the global economy—The Logistics Performance Index and its indicators*. Available at http://www.worldbank.org/content/dam/Worldbank/document/Trade/LPI2014.pdf

World Bank. (2016). *Global rankings 2016*. https://lpi.worldbank.org/international/global

World Economic Forum. (2015). *The global competitiveness report 2015–2016*. Available at http://www3.weforum.org/docs/gcr/2015-2016/Global_Competitiveness_Report_2015-2016.pdf

Blogs

Kutarna, C. (2016, June 8). *There's never been a better time to be alive. So why the globalization backlash?* World Economic Forum. https://www.weforum.org/agenda/2016/07/there-s-never-been-a-better-time-to-be-alive-so-why-the-globalization-backlash

Lehmacher, W. (2015, June 12). *The next big thing in the sharing economy?* World Economic Forum Agenda. Accessed February 11, 2016, from http://www.weforum.org/agenda/2015/06/the-next-big-thing-in-the-sharing-economy

Lehmacher, W. (2016, July 20). *How smart packaging can save lives*. World Economic Forum. https://www.weforum.org/agenda/2015/07/how-smart-packaging-can-save-lives

Lehmacher, W. (2016, September 28). *Wildlife crime: A $23 billion trade that's destroying our planet*. https://www.weforum.org/agenda/2016/09/fighting-illegal-wildlife-and-forest-trade/

About the Author

Wolfgang Lehmacher is the Director and Head of Supply Chain and Transport Industries at the World Economic Forum. The International Organization for Public-Private Cooperation is committed to improving the state of the world and engages the foremost political, business, and other leaders of society to shape global, regional, and industry agendas. Lehmacher holds a Bachelor's degree in Business Administration from the German Foreign Trade & Logistics Academy (DAV). While working in various renowned companies based in the express, post, and logistics sector across Europe and the world over the past 25 years, he has acquired a reputation as expert in global supply chain and logistics.

Prior to joining the Forum, Lehmacher was partner and Managing Director for China and India at CVA, a global strategy firm focusing on board advisory of blue-chip enterprises. In his position, Lehmacher headed the global transportation and logistics practice. He also held a range of management positions; Lehmacher was President and CEO of GeoPost Intercontinental, the global expansion vehicle of France's La Poste. For more than 10 years, Lehmacher led the international expansion of La Poste Groupe, first in Europe and then across the globe.

Lehmacher is the author of various publications including *How Logistics Shapes Our Lives* and *Logistics in the Light of Urbanization*. With his expertise in supply chain management, global logistics, and the express parcel delivery business, Lehmacher is a frequent speaker at the World Economic Forum, the Boao Forum in China, leading supply chain conferences in the world, and the Centre of Transportation and Logistics at MIT.

Index

CPSIA information can be obtained
at www.ICGtesting.com
Printed in the USA
LVHW080825080121
676002LV00004B/19